Adrian Wallwork

CVs, Resumes, and LinkedIn

A Guide to Professional English

 Springer

Adrian Wallwork
Pisa
Italy

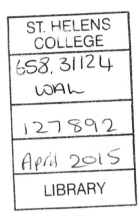
ISBN 978-1-4939-0646-8   ISBN 978-1-4939-0647-5 (eBook)
DOI 10.1007/978-1-4939-0647-5
Springer New York Heidelberg Dordrecht London

Library of Congress Control Number: 2014939420

Printed on acid-free paper

Springer is part of Springer Science+Business Media (www.springer.com)

INTRODUCTION

Who is this book for?

The book is intended for both native and non-native speakers of English. It focuses mainly on graduates and PhD students, and also young people who are already in employment and are looking for a new job. It is not intended for people who already hold managerial positions.

In the Contents page, subsections that are only relevant to:

- non-native speakers are marked with this symbol: *
- those looking for a job in academia or research are marked: §

How is this book organized?

The book is structured as a series of FAQs (frequently asked questions) with answers. The first three chapters outline:

- the quality of a good CV or resume
- how recruiters and HR people make their judgements
- whether using a template is a good idea
- how to write dates

Chapters 4–10 examine each part of a CV from your name and photo, to your personal interests and references. At the end of each of these chapters is a subsection saying in what ways, if any, a resume differs from a CV.

Chapters 11–14 regard how to write a reference letter, a cover letter, a bio, and a LinkedIn profile.

Every chapter ends with a summary.

Section 15.9 presents a possible template for a CV. Downloadable templates can be found at e4ac.com under 'CVs'.

Is this a book of guidelines or a book of rules?

Guidelines, not rules.

The book is based on interviews with recruiters and HR managers, and an analysis of hundreds of CVs from around 40 different countries.

The result is a series of guidelines on how I think a good CV and cover letter should look, they are not objective rules. Inevitably, you may not agree with all the suggestions, and are thus totally free to ignore them.

I have tried to be objective and to avoid all the possible pitfalls of writing from the perspective of a white British male. I thus hope that you will find nothing in this book that will offend you in any way. Please write to me if you feel there is anything that needs changing (adrian.wallwork@gmail.com).

Terminology used in this book

CV (also written *curriculum vitae*)

A reverse chronology listing your education, work experience, skills and interests. Generally two pages long, and typically used in all Anglo countries apart from the US and Canada.

Resume (also written *résumé*)

A brief summary of your achievements and skills, not necessarily in reverse chronological order, and generally not as comprehensive as a CV. Generally one page long, and typically used in the US and Canada.

Recruiter

Someone who works for an agency that finds potential candidates whose CVs and resumes are then submitted to the agency's clients.

Human resources (HR) manager

The person in an organization who deals with staff in general, and specifically recruitment and employment.

Hiring manager

The person responsible for deciding who to employ. This position may be held by the human resources manager, or vice versa.

For the sake of simplicity, although a CV and resume are not exactly the same (see 1.4), I will generally just use the term CV. And although a recruiter, HR manager and hiring manager do different jobs, I will often use these terms indiscriminately.

How dates are used in this book

CVs are full of dates of when you started and finished an activity. For the purposes of this book, I am imagining that we are now in 2030. So unless you are reading this book in 2030, most dates will appear to be in the future.

Examples used in this book

All the examples used in this book have been taken from real CVs, cover letters, reference letters etc. The only things that have been changed are personal details, dates and layout / font.

I use *he* or *she* at random to refer to the candidate who produced the CV or cover letter.

Other books in this series

There are currently five other books in this *Guides to Professional English* series.

Email and Commercial Correspondence
http://www.springer.com/978-1-4939-0634-5/

User Guides, Manuals, and Technical Writing
http://www.springer.com/978-1-4939-0640-6/

Meetings, Negotiations, and Socializing
http://www.springer.com/978-1-4939-0631-4/

Presentations, Demos, and Training Sessions
http://www.springer.com/978-1-4939-0643-7/

Telephone and Helpdesk Skills
http://www.springer.com/978-1-4939-0637-6/

All the above books are intended for people working in industry rather than academia. There is also a parallel series of books covering similar skills for those in academia:

English for Presentations at International Conferences
http://www.springer.com/978-1-4419-6590-5/

English for Writing Research Papers
http://www.springer.com/978-1-4419-7921-6/

English for Academic Correspondence and Socializing
http://www.springer.com/978-1-4419-9400-4/

English for Research: Usage, Style, and Grammar
http://www.springer.com/978-1-4614-1592-3/

Acknowledgements

Big thanks to Anna Southern and Philippa Holme for editing the manuscript. Also, thanks to my students, fellow teachers, friends and family who kindly allowed me to use extracts from their CVs, cover letters, personal statements, reference letters, etc.

Special thanks to Joanna Andronikou, Celine Angbeletchy, Kamran Baheri, Matteo Borzoni, Lisa Caturegli, Chengcheng Yang, Matthew Fletcher, Sarah Macchi, Leonardo Magneschi, Maral Mahdad, Sara Giovanna Mauro, Patrick Mukala, Daniela Pianezzi, Lena dal.

I would also like to thank Philippe Tissot for allowing me to include extracts from the Europass template (http://europass.cedefop.europa.eu/en/home).

Contents

1 THE QUALITIES OF A GOOD CV AND RESUME

1.1 What is the purpose of a CV?

The aim of your CV or resume is to encourage a recruiter to contact you regarding a possible job.

Write your CV from the point of view of the person who will receive it and examine it, i.e. a recruiter in an agency, an HR person in a company or research institute, a professor or fellow researcher in a research team.

This means you should:

- use a format that will be familiar to the reader (i.e. a standard template, which you can modify where appropriate) rather than a format that you have designed totally by yourself. A standard format is easier to navigate for the reader – he / she knows exactly where to look in order to find what he / she is interested in

- only include details that are relevant to the job you are looking for

- clearly highlight your skills and qualifications

- be honest, accurate and as objective as possible

A CV is thus not an opportunity for you to:

- write every single detail of your career history, education history and personal history

- experiment with your design skills

For an ironic persepctive on writing CVs, see the Polish poet Wisława Szymborska-Włodek's poem 'Writing a Curriculum Vitae' – just type in her name and the title of her poem into your seach engine.

A. Wallwork, *CVs, Resumes, and LinkedIn,*
Guides to Professional English, DOI 10.1007/978-1-4939-0647-5_1,
© Springer Science+Business Media New York 2014

1.2 How many pages should a CV be?

Two.

There seems to be a general consensus that two pages is the maximum. Your ability to be concise and to only highlight your most relevant skills and qualifications is revealed through your CV. If your CV is more than two pages, the HR person may think 'this person is unable to express themself clearly and concisely; this is not the kind of person I want working in my company'.

However, if you are looking for certain high-level or high-profile jobs in academia, it may be appropriate to use more pages so that you have space to describe the projects you have been involved in. If you have a lot of publications (see 8.10), put them in a separate document or just have a link to your website where the professor (or whoever is dealing with your application) can download them if he / she wishes.

1.3   What is the typical order of information in a CV?

The most common order is outlined below. The numbers in brackets refer to the chapters of the book where these are dealt with individually.

1. name (4)

2. personal details (4)

3. objective / personal statement / executive summary (6)

4. education (7)

5. work experience (8)

6. skills (9)

7. personal interests (10)

8. publications (8.9, 8.10)

9. references (11)

There are of course variations in this order. Some people put point 3 before point 2. If you are an academic you may put your publications (point 8) into a separate document. If you are not an academic and have work experience, then you will probably put your work experience before your education. The skills section may be divided into subsections (e.g. technical, language). Not everyone mentions their personal interests though I would argue that these are essential. Not everyone puts references.

If you are not applying for a job in academia, then your publications will be of little or no interest to the recruiter.

Not all templates will include all the nine points above. For instance, at the time of writing, the Europass (2.10) has no specific section for personal interests.

1.4 What is the difference between a resume and a CV?

The word *résumé* is French and means 'summary'. It is a concise overview, generally just on one page, of your objectives and main achievements.

It covers the same areas (points 1–8 in the previous subsection) as a CV, but presents them in a different way. A CV is more like a technical description of a product (i.e. the candidate), whereas a resume is more like a sales brochure – though a good CV will also try to sell the candidate.

For an example of a CV and a resume, see 15.9 and 15.10.

1.5   What are companies really looking for?
And research institutes?

There is no real difference in the requirements in industry and in research, though in research your academic qualifications are likely to have more weight.

Both companies and institutes clearly want to see evidence on your CV that you are qualified in terms of both education and work experience for the position that they have open. But they also want evidence that you:

- have a strong work ethic and that you work to deadlines (even under stress)

- can work in a team and are easy to get along with

- are both proactive and flexible

- have the technical, emotional and analytical skills for problem solving

- can give effective presentations

- have good communication skills

- can write reports and other kinds of documents

- are enthusiastic and passionate about what you do

- are professional, reliable, well mannered and appropriately dressed

- would fit in well with the company / institute – both in terms of the environment and the core values

You need to inject each section of your CV with evidence that you have the above attributes. The following chapters will tell you how.

Here is an extract from an email from an HR manager to a recruiting agency:

> The candidate should have:
> - Excellent communication skills
> - Analytical thinking – be able to build patterns from raw data
> - Be able to write succinct reports in English to tight deadlines
> - Be a go getter, self motivator, who can work fairly independently

If you were applying for the post indicated in the email above, your CV would need to demonstrate that you have such qualities.

For example, by writing that you have had 12 research papers accepted at international conferences you are indicating that: 1) you can write technical documents; 2) you have experience in presenting your work; 3) your English is probably of a high standard.

You should not state directly, either in your CV or cover letter, that 'I have good communication skills' as such skills are subjective and difficult for the recruiter to evaluate. Instead the recruiter should be able to understand that you have these skills from the evidence that you provide in your education, work experience and personal interest sections (see 9.7).

1.6 Is it a good idea to send the same CV to different companies / institutes?

No.

You need to tailor (customize) your CV for the specific post you are applying for.

You could start by drafting a CV that contains everything that anyone could possibly find relevant and interesting. This could require several pages. You then adapt this draft CV to make it read and look as if it was specifically written for that particular company or institute. Adaptation consists of:

• deleting anything that is not strictly relevant. This does not mean removing whole parts from your Education and Work Experience sections, but simply removing elements that are not relevant in any way to that job. This means that you can highlight the key qualifications that you could bring to the post you are applying for.

• modifying the text to make sure that it includes evidence of the skills that you will need for the post you have applied for

• changing the layout and/or font so that it reflects the same graphic style as the company or institute where you want to work

1.7 How many CVs do recruiters receive a day? How quickly do they read them?

Recruiters and HR staff personnel receive 100–400 + CVs a day if they have posted a job on the open market.

Google receive between 1,000 resumes per day. Many hires result from these CVs, but only after comprehensive reference checks.

Although some recruiters spend around two to four minutes on reading CVs, many will reject a CV after only six seconds.

1.8 How do recruiters filter CVs?

Most recruiters will require an electronic version of your CV. This enables them to use application-tracking systems that scan your CV for key words (see 6.4, 8.3, 8.4 and 14.5 for how to insert key words into your CV).

After this initial screening, the recruiter will be left with a limited number of CVs. They will then look at each CV individually, and ask themselves the following questions:

1. is this CV laid out in a way I am familiar with?

2. after six seconds can I get a clear picture of who this person is and what they can offer my company? does the person have the relevant qualifications?

3. does this CV look as if it was specifically written, or at least tailored, for my company?

4. is this person credible?

5. is this CV reasonably short (i.e. max two pages)?

If the answers to at least two of the above questions are affirmative, they put the CV aside for further consideration, otherwise they trash it.

To fulfill Criteria 1 (layout), you need to:

- use or adapt a standard format / layout, i.e. a format that the recruiter will have seen hundreds of times before and will thus be very familiar with, rather than a format that is completely new for him / her and which will thus be more difficult to navigate

- make good use of headings

- use plenty of white space

- ensure that you have not tried to include further information simply by reducing the font size (10 pt should be the minimum)

For Criteria 2 (qualifications) and 3 (tailored), you should:

- find out everything you can about your chosen research institute or company, and see how you can match the kind of jobs they have on offer (if you are writing cold, i.e. not in response to an ad) or the specific job (if you are responding to an ad)

- highlight in your descriptions of what you have done at each stage of your career (both academic and work), those skills that you learned that would be particular relevant for the job you want

1.8   How do recruiters filter CVs? (cont.)

For Criteria 4 (credibility)

- be honest about your achievements (see 1.11 and 8.5)

- don't make any spelling mistakes. Just one spelling mistake is enough for your CV to be rejected. Why? Because if you did not take the time to check your CV, this probably means you are the type of person who does not check their work in general.

Finally, for Criteria 5 (conciseness) you should:

- consider your CVs like a short abstract or summary. Be concise, short and clear. Cut all redundancy

- put the most important (and most recent) information first

- be relevant: tell recruiters only what THEY need to know, not everything that YOU know

- give maximum importance to what makes you different: sell yourself

1.9   Do all recruiters and HR people read a CV in the same way?

No. Most will start at the top and work down. However, some may start at the bottom and work up. This means that wherever the reader chooses to begin reading, what they read must be perfectly clear and make perfect sense. So don't just spend most of your time preparing a fantastic beginning to your CV, ensure that each part is perfect – including the section on Personal Interests (see Chapter 10), which is where some HR people may start.

1.10 How can I maximize the chances that my CV will not be rejected after six seconds?

As with all kinds of writing, you need to imagine that you are the reader, in this case a recruiter. Don't think of ways to impress the recruiter. Instead think: how can I make my CV easy and quick to read? This means:

- understanding and facilitating the priorities of the recruiter (i.e. their top priority is to find the right candidate in the minimum time possible with the minimum effort)

- avoiding big chunks of text by inserting white space

- avoiding unnecessary information (e.g. if you are a PhD student applying for a job in industry, the recruiter will probably not appreciate seeing the long list of papers that you have published)

1.11 Do I need to be honest?

Yes.

CVs these days are checked for detail using special software. Any discrepancies will be identified immediately. In any case, experienced recruiters can spot a lie very quickly.

Even just one thing on your CV that is proved wrong during an interview may make the interviewer think that the rest of the CV is not true either.

If you make claims on your CV which during the interview turn out to be unsubstantiated you also risk the company informing the recruiting agency, who may then remove you from their database.

1.12 Will recruiters access my Facebook account?

Possibly.

The kind of information – both in the form of text and images – that you post on the Internet is a clear indicator of your personality and of your social behavior and communication style. Also, HR people have a natural human curiosity about potential candidates and will seek out information that is not contained in a CV.

A survey conducted by UK Job Forecast found that the majority of HR people use the web as part of their strategy and will screen candidates by checking any information about them on personal websites, LinkedIn, Facebook, Twitter etc. Over 60 % of employers questioned by CareerBuilder.com rejected candidates on the basis of information that their recruiters had discovered online.

So ensure that you limit who can access your pages. Alternatively, do not post anything online that you would not want to be read or seen by an HR person (e.g. compromising photographs!)

1.13   Should I consider a video CV?

Probably not. Unless, perhaps, you are looking for a job in media or advertising.

At the time of writing this book, video CVs are not common for most types of jobs. The risk is that if you only send a video CV, no one will look at it. Moreover, it is a little more difficult to skim through a video than a written document, thus you are asking for a greater effort on the part of HR.

When a recruiter looks at a CV, he / she can concentrate almost exclusively on the information contained in it. A video CV can be watched from many points of view – not just content, but also the means of presentation (much more so than in a written CV). Your clothes, your voice, your behavior could all distract from what you are saying about yourself and your achievements. Thus the first impression that the recruiter gets may be based on superficialities such as the way you move, your tone, and your smile or lack of it. Such impressions are better left to the face-to-face interview.

However, if video CVs are the norm in your intended work field, then check out other people's video CVs. There are plenty available on YouTube and also on professional recruitment sites.

Watch them and decide what you think works well and what you should avoid.

The ones that often work the best tend to be the simplest. They:

- have no distractors, so that what the HR person sees is much the same as they would see in a good quality Skype call, i.e. you sitting down (preferably in front of a white background) without performing any particular actions
- are shot in one session, i.e. not in a series of different locations at different times

A major issue is your skills in the English language. If you are a non-native speaker:

- ensure you have your script (i.e. what you say in your video) corrected by a native speaker
- use short sentences (long sentences are more difficult to say)
- only use words that you can pronounce correctly
- enunciate clearly and do not speak too fast

1.14   Is it a good idea to have my CV on my personal website?

Yes.

If someone takes the trouble to access your personal website, they will probably also be motivated to read / download your CV.

In this case, you need to write your CV so that it will appeal to as many different types of potential employer as possible within the fields that you are looking for. Or if you are only interested in one specific area, then your online CV should reflect this area.

1.15   I am looking for my first job. Before writing my CV, what questions should I ask myself?

If this is the first job that you are applying for, consider the following:

- am I more interested in a career where I can use my skills or one which will satisfy my interests?

- how would I describe myself in one sentence?

- what are my greatest skills and how might they match the job I am looking for?

- what are my major accomplishments? how might these be relevant for a particular job?

- do I like working independently or as part of a team? would I make a good team leader?

- do I mind (enjoy) working long hours? how well do I deal with deadlines?

- what are the most important factors I am looking for in my ideal job?

Your answers to these questions should help you first decide what kind of job you would like, and secondly help you to decide the content of your CV.

1.16 How much time and effort should I spend on writing my CV?

Your CV is probably one of the most important documents that you will ever write in your life. It is your passport to a job. It is worth spending time and money to make it look 100% professional.

Look at the extracts below from a CV. Based purely on a first impression and irrespective of the candidate's field of work, would you be interested in offering this person a job? How much time and attention do you think the candidate devoted to his CV?

Curriculum Vitae et studiorum of Dario Monte

Place of birth	Pisa
Date of birth	21th august 1996
Home address	C.so italia 177
Home phone	+390508869872635
E-mail	supermariomonte@hotmail.com

EDUCATION AND JOB EXPERIENCE

18/jul/2025	Graduation in Biology cum laude at University of Pisa
11/sep/2025	Qualification as professional Biolgist
1/dec/2026	begin PhD in Molecular Technology at University of Pisa

.....

LANGUAGES

Good knowledge of spoken and written english language, also in scientific field. I attend language course at the University of Pisa.

1.16 How much time and effort should I spend on writing my CV? (cont.)

The above CV will not give a good impression, for many reasons:

- the heading 'Curriculum Vitae et studiorum of Dario Monte' is not standard. A more conventional is simply to write the name 'Dario Monte'

- the personal details take up more space than necessary and also contain a mistake (i.e. 21th august – it should be 21 August or 21st August)

- the email address is not professional

- the term 'job experience' should be 'work experience'

- the dates are all written incorrectly and the months should all begin with an initial capital letter

- the use of bold in the first qualification makes no sense

- the word 'biologist' is mispelled

- what is the qualification as a professional biologist? – outside Italy this title is unknown

- the grammar is incorrect in 'begin PhD' and in 'at University', (it should be *began Phd ... at the University ...*) and in any case why the beginning date?

- in the language section Dario makes a series of grammatical and capitalization mistakes, which indicate that, contrary to what he claims, he does not have a good knowledge of English

Dario is a highly intelligent and articulate biologist. Yet he totally failed to understand the importance of his CV.

Dario sent his CV to a recruiting agency and received this reply from the agency:

> To be completely honest I think your CV lets you down. If you are managing to be successful as a freelancer then I assume you must have good technical skills, but I am afraid this does not come across in your CV ... I realise you may like to keep your CV brief, however you need to include more details about what you actually do. ... I don't think your written English is quite up to the level that our clients would expect.... You may like to consider asking a professional to re-write your CV. ... I hope you don't mind me being so blunt.

1.16 How much time and effort should I spend on writing my CV? (cont.)

I would like to underline that this email is completely genuine, and has not been written for the purposes of this book. The problem is that Dario:

- has not taken enough time to write an acceptable CV – there is not enough detail

- probably has the right skills for the kind of job he wants, but these skills are not revealed or highlighted in his CV

- is not attentive to detail – he has not checked his English

- evidently thought that his CV was acceptable

Chapters 2–10 explain how to write and layout an effective CV.

Summary: The Qualities of a Good CV / Resume

➢ Your CV / resume is one of the most important documents you will ever write.

➢ Before writing your CV, research your chosen organization and find out what they expect to see in a CV.

➢ It should be designed exclusively to get you a job.

➢ Leave aside your own personal opinions of how a CV should be written and laid out, stick to what will be familiar for the reader.

➢ Tailor your CV specifically to your chosen organization.

➢ If it doesn't attract attention within 2–3 seconds, it will not be read in full.

➢ Write from the hirer's / employer's point of view.

➢ Be honest, concise, accurate and factual.

➢ CV: two pages max. Resume: one page max.

2 TEMPLATES

2.1 What is a template? Should I use one?

A template is a document that shows a layout and sample content. The template puts the information required (i.e. points 1–9 in 1.3) into a specific order.

For good clear templates see See 15.9 and 15.10. Downloadable templates can be found at e4ac.com under 'CVs'.

Given that your aim is to facilitate the work of the recruiter who reads your CV, it makes sense to use a template, or at least a format, that a recruiter will be familiar with. If necessary you can modify the format of your CV (see 2.14) to make your CV stand out from the hundreds of others that the recruiter will be reading.

Perhaps the three most standard templates are the various models provided by Microsoft Word, Europass (see 2.10) and LinkedIn (see Chapter 14). In fact, the popularity and importance of LinkedIn mean that CVs are very likely to look more and more like a LinkedIn profile, in terms of both layout and content.

You have to balance your aim of making your CV stand out from the rest, with the recruiter's need to find information about you as quickly and as easily as possible.

In some cases your potential employer will provide you with their own template to fill in. If this is the case, follow their instructions carefully. Do not try to adapt their template to fit your wishes. Alternatively, you may be requested to use a particular standard template.

2.2 What are the advantages of using a recognized standard template?

If you use a good template, it will make it easy for:

- you to compile your CV: the instructions to the template will tell you what information to include, where to put it, and in what order to put it

- recruiters to find the information they want and to make quick comparisons between you and other candidates

The best advice that you can get for writing a CV for a specific company or institute, is to find someone who already works there and look at their CV or alternatively ask them to look at yours.

A. Wallwork, *CVs, Resumes, and LinkedIn,*
Guides to Professional English, DOI 10.1007/978-1-4939-0647-5_2,
© Springer Science+Business Media New York 2014

2.3 Isn't the information contained in my CV more important than the layout?

No, not initially.

A reader's eye is drawn towards white space and initial capital letters. This means that we focus more on the beginnings and ends of sentences, than we do on the middle of sentences. It also means that CVs that have big blocks of dense text tend to be read with less interest than those where the content has been divided up into short blocks of text.

Recruiters' eyes tend to focus on the left hand side of the page. This is probably due to the fact that in standard well laid out CVs, the dates and key words (i.e. job positions, names of companies and universities) tend to be found on the left.

So, your CV may be quickly discarded if:

- key achievements are hidden within a big blocks of texts
- you deviate from the standard presentation, i.e. if your dates and key words do not appear on the left

The way your CV is laid out is thus crucial if you want a recruiter to look at it for more than a couple of seconds.

2.4   I want to be different. Should I create my own layout and style?

Probably not.

To understand why, try this experiment. Find five or six examples of CVs from friends, family or on the Internet. Do not include any CVs that follow a standard template (e.g. the Europass).

Look at each CV for a maximum of six seconds. Which ones do you like and not like? Why? What impression of the candidate do you get? Think about:

- how pleasing the CV looks

- what order the information is presented

- how easy it would be for recruiters to find the key information they are interested in

Then, imagine you are the HR person, and that these CVs are just five of the 250 that you have received for the same job. What would be your main problem with deciding which candidates to reject and which to interview?

You will probably notice:

- the incredible variety of presentation, layout and formatting styles

- that even the name of the candidate does not appear in the same place in each CV

- that the order of information is not the same (some begin with work experience, others with academic experience, and others with a personal objective)

- the massive difference in the way the candidates present their personal details and the abbreviations they use, not all of which will be familiar to all recruiters

- the different headings for the same kind of activity (e.g. work experience, professional experience, employment history)

What is also interesting is that presumably the people who wrote these CVs were satisfied with what they had produced.

So what is the net result for the recruiter who is faced with a massive variety of formats? Answer: confusion.

The recruiter has to work extra hard to find the information that he / she wants and to be able to compare the same information across several CVs.

If you were the recruiter, would you not prefer to receive the same information in the same way from all the candidates? Then you could

2.4   I want to be different. Should I create my own layout and style? (cont.)

focus primarily on comparing the experiences of the candidates rather than wasting time having to actually identify this experience and being distracted by different layouts.

So if you decide to be creative and to produce your own original CV template, you need to be aware that it may be detrimental to the chances of your CV being read and thus of you ever being invited to an interview. This is true in the majority of areas of both industry and research, possible exceptions may be in media and advertising, where having a creative CV may indicate a creative mind.

2.5   Should I use color?

Typically people use color:

- to highlight headings (e.g. a shade of blue) or particular achievements
- in their photograph

But it may be better to use shades of grey. If the recruiter prints the CV in black and white much of the color impact will be lost.

Your photo (see Chapter 5) will probably also look more professional if it is in black and white. If the HR person wants to see you in color then he / she can go to your LinkedIn page (or Facebook!).

2.6   What about the logos of the institutes and companies I have worked for?

You may think that your CV will look more visually appealing if you insert logos. It may look visually appealing to you, but it will simply be distracting for your reader. Keep the format as simple as possible, without logos.

2.7   What are the best fonts to use? And what size?

Among the clearest fonts to read are Arial, Calibri and Verdana.

You might also consider using the same font as is used by the company or institute where you are sending your CV—it gives the impression that you already work there!

Use between 10 pt and 11 pt. Anything smaller is difficult to read and will look as if you have tried to include too much text, rather than finding ways to be more concise.

2.8 What about spacing between lines, paragraphs and sections? And bullets?

The reader's eye is attracted by white space. Use more white space between sections than between paragraphs within a section. Only use bullets if strictly necessary. Restrict the use of all caps (i.e. words written only in capital letters).

Compare these two versions—which is easier to read? The font in both cases is Calibri, 10 point.

VERSION 1: NOT SPACED, MIX OF ALL CAPS AND LOWER CASE, BULLETS

LANGUAGES
- Chinese mother tongue
- English: fluent (spoken and written)

AWARDS AND HONORS
- Young Scientist award, *POLYCHAR 19—World Forum on Advanced Materials*, Nepal, 2019.
- Best Poster Award, *Fluoropolymer 2018*, Mèze, France.
- Excellent Graduate of Shanghai, ECUST, China, 2017.

References
- Prof. Giulia Gestri (in whose lab I did more than 4 years of research work), Department of Chemistry and Industrial Chemistry, University of Pisa. gestri@dcci.unipi.it

2.8 What about spacing between lines, paragraphs and sections? And bullets? (cont.)

VERSION 2: SPACED (6 PT BETWEEN SECTIONS, 2 PT WITHIN SECTIONS), NO CAPS, NO BULLETS

Languages
Chinese: mother tongue; English: fluent (spoken and written)

Awards and Honors
Young Scientist award, POLYCHAR 19 – World Forum on Advanced Materials, Nepal, 2019.

Best Poster Award, Fluoropolymer 2018, Mèze, France.

Excellent Graduate of Shanghai, ECUST, China, 2017.

References
Prof. Giulia Gestri (in whose lab I did more than 4 years of research work), Department of Chemistry and Industrial Chemistry, University of Pisa. gestri@dcci.unipi.it

Version 2 occupies marginally more space, but is cleaner and much easier to read. It is worth showing your CV in different formats to as many people as possible, then see if you can reach some consensus as to which is easiest to read.

2.9 What is the Europass?

In Europe, one of the most commonly used templates is the Europass Curriculum Vitae, which you can download from the Europass site.

http://europass.cedefop.europa.eu/en/documents/curriculum-vitae/templates-instructions

If you opt for the Europass, make sure you use the most recent version as the template is being constantly streamlined.

Having a professional template helps to ensure that you provide all the information that is typically required by HR and recruiters. However, note the following possible problems with the Europass template:

- there is no Executive Summary. See 6.5 to learn why an Executive Summary is a key element to your CV

- there is a separate section for Communication Skills. See 9.7 for why it is probably best not to have such a section

- there is no section for Personal Interests. See Chapter 10 to understand why HR people might want to see what extra curricula activities you are involved in

- there is no References section. See Chapter 11 to learn why references are important.

Note: since the publication of this book the Europass template may have changed, and thus the above points may no longer be valid.

Standard templates usually have instructions on how to compile them, and such instructions generally offer very useful advice. Here is an extract from the Europass notes relating to Work Experience:

– if you are applying for your first job, do not forget to mention work placements during training which provide evidence of initial contact with the world of work;

– if your work experience is still limited (because you have just left school or university), describe your education and training first (to invert the order of the two headings, use the 'copy / paste' command in your word processing software); highlight work placements during training (see online examples);

– for the sake of brevity, focus on the work experience that gives added weight to your application. Do not overlook experience which may be an asset even though it is not directly related to the profile of the job for which you are applying (e.g., time spent abroad, work bringing you into contact with the public, etc.);

2.9   What is the Europass? (cont.)

The printed version of a Europass CV looks very professional, and this is clearly an advantage over a homemade template. It is also an officially recognized CV, which may help to give you extra credibility.

A standard template will probably also be familiar to the recruiter who reads it. This will enable them to find the information they need quickly as they know exactly where to look for it. It will also enable them to make quick comparisons between your CV and those of other candidates that also wrote their CV using the same template.

For more templates, go to my website e4ac.com and look under the menu for 'CVs'.

2.10 With my chosen template my CV has become three pages long, what can I do?

The example below highlights that some templates can waste a lot of space.

Education and training

Dates	2028–2031
Title of qualification awarded	PhD
Principal subjects / occupational skills covered	Thesis Title: 'Young People in the Construction of the Virtual University', empirical research that directly contributes to debates on e-learning.
Name and type of organisation providing education and training	Brunel University, London, UK Funded by an Economic and Social Research Council Award
Level in national or international classification	ISCED 6
Dates	2024–2027
Title of qualification awarded	Bachelor of Science in Sociology and Psychology
Principal subjects / occupational skills covered	- Sociology of Risk, Sociology of Scientific Knowledge / Information Society; E-learning and Psychology; Research Methods.
Name and type of organisation providing education and training	Brunel University, London, UK.
Level in national or international classification	ISCED 5

In the example above the last two items in the left hand column each cover two lines, whereas the corresponding right hand column only takes up one line. There is also a lot of repetition in the left hand column. This is wasted space, particularly if you have a lot to write. If you have need to put three or four (or more) activities under *Education and training*, your CV is likely to go over the recommended two-page limit.

2.10 With my chosen template my CV has become three pages long, what can I do? (cont.)

The solution is simply to adapt the template as shown below:

Education and training

2028–2031	PhD (ISCED 6), funded by an Economic and Social Research Council Award at Brunel University, London, UK
	Thesis Title: 'Young People in the Construction of the Virtual University', empirical research that directly contributes to debates on e-learning.
2024–2027	Bachelor of Science in Sociology and Psychology (ISCED 5), Brunel University, London, UK.
	Principal subjects / occupational skills covered: Sociology of Risk, Sociology of Scientific Knowledge / Information Society; E-learning and Psychology; Research Methods.

The modified example above has reduced the original template from 22 lines to 12 lines—nearly 50 %. The headings in the original left-hand column have been integrated into the new right-hand column. The rest of the text in the original left-hand column has been deleted, but no information has been lost.

Also, the modified version uses Arial rather than Arial Narrow which makes it easier to read.

These kinds of modifications should help you to reduce a 3–4 page CV based on a standard template into a 2-page CV.

2.11 I have been requested to use a particular template, can I customize it?

No.

The reason why certain employers request a specific template is so that all the CVs they receive will look the same. They can thus read and filter them quickly. If you customize their template, you make the HR person's job more difficult and risk irritating them.

Those recruiters who request, for example, the Europass, will know not to even look at the left-hand column and instead will just skim down the right-hand column.

2.12 I already have my CV in the requested template, but in my own language. What do I need to be careful of?

The risk is that when you convert it into English, you will forget to translate everything. This is an incredibly common problem. For instance, you may forget to translate:

- headers and footers
- text in the left-hand column
- names of qualifications and institutes

Another problem is that you might decide to cut and paste your original CV into the English format. In this case, you need to check carefully to ensure that nothing of your original language remains and that the font and layout do not change as a result of pasting.

If you have never written a CV before and you are intending to find work in an international environment, it makes sense to produce it in English. Even if you later apply for a job in your own country, you will find that most big corporations or research institutes will accept (or even request) your CV to be in English.

2.13 I have decided to use a template. What can I customize?

If you have decided yourself to use a particular template, and your potential employer has made no specific requests regarding the format of your CV, then you can customize as you wish. However the final result must match the standard norms, e.g. the usual order that information is presented.

Of course, there may be more than one 'usual' order'. But some sections tend always to appear in the same place. This means that you should not put your personal details at the end of the CV, given that most recruiters will expect to find it at the beginning.

HR people will be able to learn a lot about you from the layout of your CV (e.g. how well organized you are, whether you are a good communicator), so be careful how you deviate from the standard.

Customizing can be done in many areas.

CHANGE THE SECTION TITLES

Instead of, for example, using the term 'Personal information' you could write 'Personal Details' (note the use of initial capital letters for both words).

CHANGE THE ORDER OF THE SECTIONS

Many templates put 'Personal Information' first followed by 'Desired employment/Occupational field'. Many candidates reverse the order of these two sections to give prominence to the type of job they are applying for.

CHANGE THE ORDER OF INFORMATION WITHIN A SECTION

In the Work experience section in the some templates, the name and address of the employer is the fourth item, you might wish to put this information in first position, particularly if you have studied at a prestigious university or done an internship in an internationally well-known company

CHANGE THE FONT

The Europass uses Arial Narrow which is a nice clear readable font that does not occupy too much space. However, some people find it a little too small to read. Alternatives could be Arial (normal), Helvetica, Verdana, and Calibri—all of which are easy to read and fairly standard. Avoid using

2.13 I have decided to use a template. What can I customize? (cont.)

strange fonts (they may get you noticed but for the wrong reasons) and never use Comic Sans (or similar fonts) as it is often associated with children.

CHANGE THE COLOR

Some templates are just in black and white. You could use grey or blue for headers. In any case stick to one font and two colors. Use bold but avoid italics (unless in titles of books, papers or theses). The idea is that your CV should look clear. The more fonts, colors and formatting types you use, the more difficult it will be to read.

ADD A NEW SECTION

You can add sections that don't appear in the standard template (e.g. personal interests, references).

2.14 What kind of information is not worth mentioning in my CV?

Unfortunately, many templates encourage you to tell the entire story of your life in full detail. The instructions to filling in the template fail to tell you that you need to prioritize the information you give, and that some information is not worth giving.

Chapters 4–9 outline what to include in each section of a CV, irrespective of what template you decide to use.

See 15.9 and 15.10 for templates for CVs and resumes.

Summary: Templates

➢ Templates are worth using. To convince yourself, look at a variety of CVs that do not follow a standard template—how easy would they be for a recruiter to compare?

➢ Templates make it easier for recruiters and HR, as the information provided by candidates follows the same pattern and is thus easier to locate.

➢ Modify your chosen template to keep it to two pages, and to make it as concise, clean and readable as possible.

➢ Never modify a template that has been specifically provided by your chosen organization.

➢ Whatever you decide:

- use a readable font (e.g. Arial, Calibri)

- no smaller than 10 pt

- clear spacing between sections

- limit the use of bold and capitals and in any case always use them for the same function

- only use color if you are convinced it will increase your chances of getting a job

- ensure that you haven't left any text in your own language

3   WRITING DATES

3.1   How should I write my date of birth?

The simplest way to write your date of birth in a CV is the day as a number, the month as a word, and the year.

11 October 2020

Visually, this is the least confusing layout for dates.

Another standard way, often used in the USA, is:

October 11, 2020

The first system is clearer because the two numbers are separate and there is no need for a comma. You can use the same system when writing the date in a letter (e.g. in your cover letter).

When you need to write the full date, don't use any other system than the two indicated above. For example, do not these formats:

October 11th, 2020 (you might write 11st or 11rd by mistake)

10.11.20 (in the US this means October 11, in most of the rest of the world it means November 10)

A. Wallwork, *CVs, Resumes, and LinkedIn,*
Guides to Professional English, DOI 10.1007/978-1-4939-0647-5_3,
© Springer Science+Business Media New York 2014

3.2   How should I write the date in the Work Experience and Education sections?

You only need to write the month in two cases.

Firstly, if it refers to a very recent date (i.e. of this year, or the end of the previous year). For example, if we are now in May 2029, you might write this range of dates to describe an internship: *Jun 2028–Mar 2029*. To learn how to abbreviate months see 3.5.

Secondly, if the period of time was very short and goes from the end of one year to the beginning of the next year:

Nov 2024–Jan 2025

Imagine that we are now in the year 2030. The extracts only show parts of each section.

Work experience

Dates	January 2029—present
Occupation or position held	Researcher (full-time contact)
Main activities and responsibilities	Research in the field of Context Aware computing: - Human Activity Monitoring and classification - Environmental monitoring (glaciology) - Energy consumption issues in battery driven devices
Dates	01 October 2028–31 December 2028
Occupation or position held	Teaching assistant
Main activities and responsibilities	Laboratory course on Operating Systems—FreeBSD

Education and training

Dates	2025–2028
Title of qualification awarded	PhD in Information Engineering
Principal subjects / occupational skills covered	- Engineering problems analysis - Software design and programming - Computer networks architecture

Here is an explanation of the dates used:

January 2029–present: given that we are now in 2030 (i.e. the year after 2029), the month (January) is quite important as it indicates that the candidate worked for the entire year of 2029, not just for part of it

01 October 2028–31 December 2028: again this is quite a recent date. The candidate has put the days as well as the months to indicate that she worked a full three months

2025–2028: given the length of the course, the day and month which you started and ended a degree are not important

3.3 What about the date of my graduation / thesis?

Just write the year. If you graduated very recently: the month and year.

3.4 Can I abbreviate years?

Abbreviating years looks strange, as highlighted by the bad example below:

12–14 Cambridge University—degree in psychology

It is better to re-arrange your layout, or cut some text, than to save space by removing the first two digits from the beginning of a year.

3.5 How should I abbreviate the months?

If for space reasons you choose to abbreviate months, just use the first three letters: Jan, Feb, Mar, Apr, May, Jun, Jul, Aug, Sep, Oct, Nov, Dec.

The above are the official abbreviations, do not invent your own. You do not need to use a period (.) at the end of the abbreviation.

Good examples:

Nov 2020–Feb 2021

May–Jun 2023

Oct 2024

3.6 How do I express a range / period of time? And what if it includes the present day?

If you don't need to mention the month, write:

2019–2021

If the month is relevant:

Nov 2024–Jan 2025

If you began something in the past and are still doing it now, use either *present* or *now*.

2014–present

2014–now

The above represent the standard. Do <u>not</u> use any of the alternatives given below:

from 2019 to 2021

since 2019

2014 / 15

2014–current

2014 to present day

2014 to date

3.7 In what sections are dates not relevant?

Dates are used to show the recruiter how your experience has been built over the years. Dates are thus crucial when describing your work experience and education.

However, when you are talking about your skills, e.g. when you did an English course or when you got your computer 'licence' (in Europe), the recruiter is probably not interested in knowing when you did the course. In any case, if you obtained the qualification many years ago and you give the exact year, then the recruiter may consider it outdated.

Don't mention dates in the Personal Interests section.

3.8 Why is there a whole chapter on how to write dates in this book—isn't it a little excessive?

The way you write dates affects the overall appearance of your CV, as you can see from the following examples from the Education section.

BAD EXAMPLE 1

| March 2015–March 2016 | Master study of Materials Science, School of Materials Science and Engineering, East China University of Science and Technology (ECUST), P.R.China. |
| October 2016–November 2022 | PhD study of Chemistry and Physics of Polymers, School of Chemistry and Chemical Engineering, Shanghai Jiao Tong University (SJTU), P.R.China. |

The non-use of abbreviations of the months means that the dates intrude into the main text. In any case, the months are irrelevant and should be deleted. Also the information is in chronological rather than reverse chronological order.

BAD EXAMPLE 2

| *School leaving certificate*: | Qualified technician at the Federal Technical Institute of Bahia at the end of the year 2010 / 11. |
| *Degree in computer science* | Graduated from the Universidade Federal do Amazonas on 19 / 12 / 14. |

The dates appear at the end of the text, this makes them hard to spot for the recruiter.

3.8 Why is there a whole chapter on how to write dates in this book—isn't it a little excessive? (cont.)

BAD EXAMPLE 3

- o Since October 2020–PhD student in Cognitive Sciences at the University of Siena, Italy.
- o M.A. in Language Science at the University Ca' Foscari of Venice, Italy, 2017.
- o 2015–2016: Erasmus student at the University of Turku, Finland.

The dates appear both at the beginning and the end.

In addition to looking messy, the problem with all three examples is that they do not conform to the standard way of writing dates. This will not be appreciated by recruiters.

GOOD EXAMPLE

Education

2016–2022	PhD in Chemistry and Physics of Polymers, School of Chemistry and Chemical Engineering, Shanghai Jiao Tong University (SJTU), P.R.China.
2015–2016	Master's in Materials Science, the Key Laboratory for Ultrafine Materials of Ministry of Education, School of Materials Science and Engineering, East China University of Science and Technology (ECUST), P.R.China.

Summary: Writing Dates

➢ Do not underestimate how writing dates in the cleanest / clearest form can radically affect the way your CV looks.

➢ Date of birth: 10 March 2003 (day as a digit, month as a full word, year in four digits).

➢ Current position: 2015-present (only put month if you began very recently).

➢ Previous position: 2011–2015 (only put months for short periods, e.g. Jan 2015–May 2015).

➢ Thesis: month+year.

➢ Abbreviations: don't abbreviate years, abbreviate months by using first three letters (Jan, Feb, Mar).

4 PERSONAL DETAILS

4.1 What should I put at the top of my CV?

The first words at the top of your CV should be your name. This will enable the recruiter to find your CV quickly.

I suggest you:

- write your name as outlined in 4.2
- use a bigger font size than the rest of the CV
- use bold
- center your name or at least put it in a prominent position

Here is an example

Caturgli Elisa

Biotechnologist

elisa.caturgli@gmail.com
0039 340 7888 3147
27/01/1986
Italian

A. Wallwork, *CVs, Resumes, and LinkedIn,*
Guides to Professional English, DOI 10.1007/978-1-4939-0647-5_4,
© Springer Science+Business Media New York 2014

4.1   What should I put at the top of my CV? (cont.)

Note how the example:

- does not contain the term *curriculum vitae* or *resume* (there can be no doubt to the reader what this document is). Instead it contains the candidate's name

- does not contain any heading. It is obvious that these are personal / contact details, so there is no need for a heading saying 'Personal Information'

- the personal details are minimal and occupy little space

- the qualification / role of the candidate (biotechnolgist) is immediately clear

Some templates use icons before the address, phone number, email address etc. Such icons are not necessary, as the meaning is clear without them.

If you include your photo, there is probably no need to specify what sex you are.

An alternative, if you decide not to put your photo (see 5.1) is:

Caturgli Elisa

Biotechnologist

27/01/1986, Italian, elisa.caturgli@gmail.com, 0039 050 314 5750
LinkedIn: linkedin.com/pub/dir/Elisa/Caturgli

The version with the photo occupies the same space as the size of the photo.

The version without the photo occupies minimal space and puts all the relevant personal details on the same line. In this case, the sex is included just in case it is not clear from the name what sex the candidate is (see 4.9).

Note how Elisa has also put her LinkedIn address. This means that the HR person or recruiter can learn further details about her (see Chapter 14 on how to write a LinkedIn profile).

4.2   How should I write my name?

In Anglo countries such as the USA and the UK, people write their names as follows:

1. given name (i.e. the name your parents gave you), also known as 'first name'

2. family name (i.e. the last name of your father / mother), also known as 'surname' or 'last name'

For example, James Bond and not Bond James.

If in your country you do not follow the same standard, then it is your decision whether you conform to the Anglo system on your English CV.

Imagine you are from, for instance, Vietnam, and your name is Bui Thanh Liem, then Bui is the family name, Thanh the middle name, and Liem the given name. I suggest you put Bui Thanh Liem at the top of your CV, i.e. following the same order as in your country. However, at the interview you could explain that your given name is Liem.

4.3   I am doing an online application. There are two separate cells in the form entitled *family name and first name*. I don't have a family name, what should I do?

In some parts of the world, people do not have a family name. For example, the Ethiopian name Yohannes Gedamu Gebre is in the order of given name, father's given name and grandfather's given name. However, given that the reader of your CV is unlikely to have this knowledge, the solution is to treat the third name as the surname when applying for a job.

4.4   My name has accents / diacretics – should I use them?

If your name contains a lot of accents and diacretics and is not a name known in Europe or the US, then it is probably simpler not to put the accents and diacretics. For example, a name such as Trần Ánh Nguyệt is probably better written as Tran Anh Nguyet.

4.5   I am Chinese, can I use my English nickname?

Some Chinese people give themselves an English nickname in order to facilitate communication with people outside China. However, on your CV you should write your official Chinese name (though obviously using Latin characters and not Chinese ones). If you use your English nickname, the recruiter might think that the English name is your real name and thus that may be your mother or father are from an Anglo country.

4.6   I am a woman from southern India. How should I write my name?

In some parts of India a woman refers to herself on her 'Indian' CV as being the 'daughter of' her father. Below is an example from the top of a CV:

K.S.Meenakshi

D/o Mr.K.Settigere, 264, Karpagam Illam, N.S.K.Street, Bethaniapuram
Madurai-625016, TamilNadu, India.

Someone who is not familiar with this convention, might think that Meenaskshi was the person's last name, and that K and S were the first letters of the person's given name. In reality Meenaskshi is this woman's first name, the K is the first letter of her father's first name, and S (for Settigere) is her father's last name. For the purposes of an international CV, the candidate would be better to write her name: Meenakshi Settigere.

There are two other problems with Meenakshi's reference to her name. First, outside India, it is unlikely that a recruiter will know what d/o stands for (daughter of). Second, without knowing what d/o stands for, a recruiter cannot know that Meenakshi is a woman (see 4.9).

4.7 What personal details do I need to include?

If your CV is going to an Anglo country, you only need to put your contact details, i.e. one phone number and one email address.

Use only one phone number and preferably a mobile number, then you will always be contactable. Also, you avoid the recruiter having to make a decision about which number to call.

Include one email address, preferably your personal email address (rather than a work address) or possibly your university email address (if you are in research).

Unless you have good reasons not to do so, you may also like to include:

- your date of birth
- your nationality

4.8 How 'professional' does my email address need to look?

Your email address reflects your level of professionalism. Avoid any of the following types of address:

lordofdarkness@yahoo.com (name of favorite rock band, movie etc)

andrew1999@hotmail.com (first name + number / date of birth)

verwhite@gmail.com (merge of first name and second name, i.e. Veronica White)

Instead, clearly differentiate your first name from your last name. Here is my address:

adrian.wallwork@gmail.com

It looks professional and no one is going to get a negative impression from it. Also it will be easy for someone to find your address within their email system. If your name is already 'taken', then try adding an extra dot after your second name and add another word or abbreviation. For example you could use a word describing your profession e.g. adrianwallwork. writer@gmail.com.

4.9 What is the legislation regarding personal details on CVs?

Equal Opportunities legislation in many countries means that you are NOT obliged to include a photo (see 5.1) or state your:

- age

- gender

- marital status, children

- race / nationality

Such legislation is extremely important and is designed to make sure that everyone has an equal chance of getting a job.

Of course, with regard to your age, the recruiter will be able to make a good guess of how old you are from the dates of your education on your CV.

Names are a particular case. If your name does not give a clear indication of what sex you are to someone who is not of your nationality, then you could decide either to include a photograph or to state your gender. This will then avoid any initial embarrassment in a phone call or face-to-face interview, when maybe the recruiter is expecting someone of the opposite sex.

Alternatively in your cover letter, you could sign yourself, for example:

Mr Andrea Rossi (if you are a man),

Ms Andrea Schmidt (if you are a woman).

4.10 What other personal details are redundant?

Given that your aim is not to waste valuable space on your CV, you do <u>not</u> need to include the following information (see also 4.9):

- your traditional postal address (either home or work) – recruiters are only likely to contact you by phone or by email

- names of other members of your family (e.g. your father's name – note: this applies to some African and Asian countries)

You also do not need to include / state:

- your fax number

- Skype, Facebook, Twitter addresses

- whether you have completed your military service

- whether you have a driving licence

4.11 What other personal details might be useful for a recruiter?

You might like to include the address of your personal website or LinkedIn profile.

4.12 Any differences in a resume?

A resume typically contains the same details as a CV, placed in the same position.

Summary: Personal Details

➢ Name: given name + family name (e.g. James Bond); don't use any nicknames.

➢ Personal details: contact details are enough (one email address and one phone number).

➢ Email: ensure you have a standard professional address that contains your real name and possibly without any numbers.

➢ Equal opportunities legislation means you are not obliged to include the following information: age, gender, marital status, nationality. However, there is nothing to stop you from including it if you feel that this will increase your chances of getting a particular post in particular company / country.

➢ Other details: if you want you can give links to your LinkedIn or Academia profiles, or your website. No other details are required.

5 THE PHOTOGRAPH

5.1 Should I include a photograph?

If the hirer expects a photograph then it makes sense to include one. Even if you don't like the idea of putting your photo on your CV you should try to satisfy the recruiter's requirements, not your own personal preferences. So if requested, include a photo.

If the hirer specifically asks you not to include a photograph then do not include one. For example, there may be Equal Opportunities legislation which prohibits them from hiring on the basis of gender or other types of discrimination.

Otherwise there are no clear guidelines to follow.

Some HR people find the photo distracting, particularly when they are totally indifferent about whether to employ a woman or a man.

However, small companies or research groups may be curious to know what you look like and how old you are. This is because you may potentially be a future colleague (or boss) of the person who reads your CV or interviews you. In such cases, if you do not include a photograph, they may try to find one on Facebook or other social network.

So if you think there is a chance that your hirer may want to see what you look like, then either include a photograph, or at least make sure there is nothing compromising on your Facebook page!

A. Wallwork, *CVs, Resumes, and LinkedIn,*
Guides to Professional English, DOI 10.1007/978-1-4939-0647-5_5,
© Springer Science+Business Media New York 2014

5.2 I have decided to include my photo. What kind of photo should I choose?

Your photo is your image. It could potentially tell the recruiter a lot about you:

- your personality - are you smiling? is it a sincere smile?

- your attitude to your appearance - the clothes you are wearing

- your level of professionality - how much trouble you have taken to choose an appropriate photo, e.g. not a photo that was clearly never intended to be put on a CV

In order not to distract the recruiter, your photo should be as 'neutral' as possible. This means:

- a professional passport type photo

- passport size

- no background

- black and white

A black and white photo tends to look more professional and also photocopies / prints better than a color photo.

5.3 I don't want to use a passport type photo. How should I choose another type of photo?

You are probably not the best judge of which photo to include on your CV. To get a subjective opinion, give some friends, or better acquaintances, a choice of five photographs of yourself. Get them to choose the one that they think would be best. You will be surprised at how much consensus there will be.

5.4 What are the qualities of a good photograph?

Look at the photos of your connections on LinkedIn and decide which ones you think would be the most appropriate for a CV. What you will probably notice is that the photos you have chosen have the following in common:

- the background is white and empty
- the candidate is in the exact center of the photo, there is not too much space above the head or below the shoulders
- he / she looks professional (smart clothes, neat hair style)
- he / she has a friendly expression (probably smiling)
- the photo would look good even if photocopied in black and white

5.5 For religious reasons I wear a headscarf. Should I wear the headscarf in my photo?

Yes.

You probably won't apply for a job in a country where headscarves are prohibited at work, so there is no reason not to wear a headscarf in your photo.

5.6 Where should I place the photograph in the CV?

It should be positioned on the left or right of the page, with your personal details on the other side i.e. it should not be centered as this gives it too much importance and also wastes a lot of space.

5.7 What other factors should I consider when choosing a photograph?

The photo should be recent and should reflect as near as possible how you will appear at your interview.

The photo should be realistic. In order to avoid any confusion at the interview, it should show you as you really are, rather than how you would ideally like to appear.

5.8   Any differences in a resume?

Most resumes do not contain a photo, but if you have space there may be no harm in putting one.

Summary: The Photograph

> ➤ You are not normally obliged to put a photo. However, a good simple black and white photo is unlikely to detract from your CV and may satisfy the curiosity of the hirer.

> ➤ Choose a black and white photo.

> ➤ Head shot only, centered on a white background.

> ➤ Make sure your hair and any visible clothes look professional.

> ➤ If possible, have a natural smile. In any case try to look friendly.

> ➤ The photo should be recent and reflect how you really look.

6 OBJECTIVES, EXECUTIVE SUMMARIES AND PERSONAL STATEMENTS

6.1 What is an Objective?

An Objective states what kind of job you would like. Typically it is used when you are not responding to a specific advert. Instead you are sending your CV in the hope that the recruiter or HR person may have a suitable job for you.

An Objective is located immediately under your personal details.

Anna Southern

anna.southern@virgilio.it, +39 340 7888 3455

Objective: Position as editor of romantic novels aimed at a young (20-30) female audience.

Note how the Objective does not require a section heading, but the word *Objective* should simply be inserted before the statement itself.

A. Wallwork, *CVs, Resumes, and LinkedIn,*
Guides to Professional English, DOI 10.1007/978-1-4939-0647-5_6,
© Springer Science+Business Media New York 2014

6.2 What should I write in my Objective?

Below are some examples.

> Objective: A position in a private industry as a technical surveyor of seismic threats.

> Objective: A position requiring expertise as a risk analyst in a company exporting to the Far East

> Objective: To obtain a full-time challenging position that offers opportunities to learn and progress while utilizing my experience in financial management.

Ensure that you do not talk about the benefits for you of working for them:

> I am interested in a position where I can further my knowledge of x and gain experience in y.

In the sentence above, the candidate implies that he is only interested in joining the lab / company so that he can improve himself and further his career. Instead, you should clearly highlight what you can offer the company, i.e. that you have some specific expertise and knowledge and that you are offering this knowledge to the lab / company for their benefit rather than solely yours:

> I am interested in a position which enables me to exploit my background in x and offer my experience of z.

In both examples above the candidate's final aim is the same (to gain experience) but the way he expresses it is totally different: in the first the focus is on him, in the second the focus is on the hirer.

You may simply wish to state what your career objective is, as in the following examples:

> A career in engineering physics with a special focus on materials science and engineering.

> Employment in the foodservice industry, particularly the healthcare sector.

> A position in teaching, specializing in helping children with learning disorders.

> A profession in veterinary medicine with emphasis on agriculture and animal production.

6.3 What are the dangers when writing an Objective? How can I avoid ambiguity?

Given that an Objective tends to be just one sentence, it is very important to put the various parts of the phrase in an unambiguous order. Can you see the ambiguity in these phrases? How could it be avoided?

1. A position as a technical surveyor of seismic threats in a private company.

2. A position offering opportunities to demonstrate expertise and progress in the field of drafting specifications for software.

3. A private industry challenging training position focusing on alternative career work style development

The ambiguities are:

1. it seems like the seismic threats will impact on private industries. Better:

   A position in a private industry as a technical surveyor of seismic threats.

2. the problem here is that the sentence is spread over two lines. It seems that the candidate wishes to demonstrate *expertise and progress*, whereas *progress* is a verb in this case (not a noun) and relates to *field* and not *expertise*. Better:

   A position offering opportunities to demonstrate expertise, and to progress in the field of drafting specifications for software.

3. this sentence reveals a typical problem: in an attempt to be concise and to use the minimum number of words possible, the candidate has written two strings of nouns which make her Objective difficult, if not impossible, to read. The solution is to use more prepositions and verbs:

   A challenging position in training for a private industry focusing on developing alternative work styles to enable staff to enhance their careers

6.4 How important is it to insert key words into my Objective?

Very.

In the example below, the candidate's Objective is too generic. There are no key words that a search engine would be able to find.

General management position utilizing extensive expertise in a major organization.

It would be better to write:

Senior management position in a Fortune 500 company utilizing 10 years' expertise in IT sales and marketing.

6.5 What is an Executive Summary?

An Executive Summary is often used when you are applying for a specific advertised job. It is a summary of who you are and enables the recruiter to get an instant idea of your qualifications and skills without needing to read the whole CV.

The secret is to highlight your unique skills and achievements, i.e. factors that will differentiate you from other candidates.

An Executive Summary is sometimes called a Personal Profile, or Career Highlights.

Like an Objective, you should place it immediately below your personal details. You do not need a heading, but you might like to make it stand out by giving it a light grey background or putting it in a box.

6.6 What is the best format - one single paragraph or a series of bullet points?

Below are three examples from academia. From a purely visual point of view, which format do you think is:

- easier to read?

- more dynamic?

- would be easier to highlight that your qualifications match the requirements of the institute or industry where you are applying for a job?

FORMAT 1 (ONE PARAGRAPH)

Five years' experience in molecular biology / genetic engineering of microalgae, focusing on fermentative metabolism and biofuels (hydrogen) in *Chlamydomonas reinhardtii*. Seven years' experience in plant adaptations to low oxygen levels; highly skilled in rice in vitro culture and transformation, gene cloning, over-expression / silencing, gene expression analyses and proteomics. Able to independently set up protocols and address related problem-solving tasks. Excellent communicative, social and presentation skills combined with strong international background. Currently in the last year of an Alexander von Humboldt postdoctoral fellowship.

Pros: takes up less space than the other formats.

Cons: not as easy to read as the two formats, difficult to pick out key information.

Conclusion: only use if short of space.

FORMAT 2 (HEADINGS RELATED TO EXPERIENCE, EXPERTISE AND INTERESTS)

Experience in syntheses of organic molecules and polymers especially fluorine-containing (meth)acrylate monomers, macromolecular initiators and macromolecules with controlled architecture.

Good knowledge of controlled / "living" radical polymerization methods e.g. ATRP, RAFT.

Future interests: Supramolecular polymers, well-architectured macromolecules by controlled polymerization, hybrid organic-inorganic nanocomposites ...

6.6 What is the best format - one single paragraph or a series of bullet points? (cont.)

Pros: easy to see key information. Allows candidate to mention what he / she would like to do in the future, which is useful if you are not responding to a specific advertisement, but are simply sending your CV to a company or institute in the hope that they might have a position open in your field.

Conclusion: perfect for academic positions.

FORMAT 3 (BULLET POINTS)

- Over 8 years of experience of **managing an Intellectual Property** department in a large research center with more than 700 research scientists, and building a portfolio of over 200 patent applications in more than 20 countries.

- First-hand experience of licensing negotiations and **successful technology commercialization.**

- Educational background in **Engineering, Management and Intellectual Property Rights.**

- **Consultation** to several universities re establishing technology transfer offices.

- More than **70 publications**, including 3 books as author or co-author, 9 peer-reviewed publications, 25 journal and newspaper articles; plus 30 conference papers.

- Teaching at more than **120 workshops on Innovation and IP Management** at universities, research centers, public and private companies.

- Creation of a **website on IP and Innovation Management.**

Pros: easy to see key information, allows candidate to show how he / she matches the requirements in the job description (the order of the bullets could follow the same order as the list of requirements in the advertisement)

Cons: takes up more space than the other formats.

Conclusion: fine if you have sufficient space.

6.6 What is the best format - one single paragraph or a series of bullet points? (cont.)

FORMAT 4 (HEADINGS RELATED TO SOFT SKILLS)

A creative and conscientious teacher of English as a foreign language. A recent Trinity Cert. (TESOL) graduate with extensive previous experience in business.

A dynamic, confident verbal and written communicator - in business and in the class room

Innovative and resourceful - an instinctive problem-solver with a flexible approach

Student / stakeholder-focused - enthusiastic and adaptable, committed to achieving results

Organised and reliable - with strong analytical and planning skills

Pros: easy to see key information. Allows candidate to highlight her soft skills (which are incredibly important in a teaching / learning environment).

Cons: they are such generic skills that they cannot be classified amongst the key words that might be picked up by the software that hirer's use to scan CVs.

Conclusion: suitable for recent graduates with little or no work experience

6.7 What tenses should I use in an Executive Summary?

The Executive Summaries in the previous subsection (6.6) highlight that in most cases you don't need verbs, so the problem of tenses does not arise. However, below is an example where the candidate has begun each bullet with a verb and has correctly used the past simple to refer to past experiences (first three bullets) and the present simple to refer to skills (last bullet).

Executive Summary

- Designed over 50 websites for 30 clients in local government
- Optimized internal search engines of existing websites for 10 clients
- Implemented basic monthly maintenance of nearly 100 websites
- Able to use all current web 4.0 technologies

6.8   How can I match my Executive Summary to the job specifications?

Let's imagine the specifications for the job (hereafter job spec) you are applying for are:

Website designer

Must be 100 % familiar with web 4.0.

At least three years' direct experience.

Excellent knowledge of search engine development.

Should be prepared to carry out also routine work e.g. website maintenance

Fluent English (both spoken and written)

The Executive Summary could be written so that it exactly matches the job spec:

Executive Summary

- Able to use all current **web 4.0** technologies
- **Four years** experience in **website design**: over 50 websites for 30 clients in local government
- Optimized internal **search engines** of existing websites for 10 clients
- Implemented **basic monthly maintenance** of nearly 100 websites
- **Fluent English** (Cambridge Proficiency, Grade A).

Note how the order of the points in the Executive summary is now the same as the order in the job spec. A recruiter's software will spot all the key words from the job spec that are in the Executive Summary. Also, a human reader will see the key words clearly as they are highlighted in bold and will be able to easily tick off all the items in the job spec.

Note also how the candidate has added her English skills (compared to the version in 6.7), as these are required in the job spec. Again, this will increase her chances of having her CV shortlisted.

6.8 How can I match my Executive Summary to the job specifications? (cont.)

In summary:

- adapt your existing executive summary to match precisely the requirements of the job spec

- put the items in your executive summary in the same order as they appear in the job spec

- add any items that are in the job spec but which were not in your original executive summary

- ensure that you insert all the key words from the job spec and highlight them in bold.

6.9 What kinds of words should I use, and what words should I avoid?

When you write your Executive Summary or Personal Statement (see 6.12), you are trying to sell yourself to the reader. However you do not want to exaggerate your abilities as otherwise you will seem less credible. So do not fill your statement with adjectives such as *amazing, best, outstanding* unless you can provide concrete evidence of such attributes.

Try to use words that will give a positive impression such as *achievement, active, evidence, experience, impact,* and *plan.*

Avoid negative words such as *bad, error, fault, hate, mistake, never, nothing,* and *problem*

6.10 How useful is an Executive Summary?

An executive summary forces you to think about what you kind of position you really want and what skills you possess to obtain such a position. You can then adapt it for your LinkedIn profile (see Chap. 14).

It is also useful for

1. a hiring manager to immediately see who you are, what you want, and how you might fit in with their hiring plans

2. a recruiting agency to paste into an email to a company who the agency thinks might be interested in you.

Below is an email from a recruiter who thinks that Carmen, who is a hiring manager in a company, might be interested in a candidate called Juri Nizik.

Dear Carmen,

I think you might be interested in Juri Nizik, a very strong candidate:

Nine years of development experience. Strong core Java/J2SE - especially in high performance multi-threaded server development. Excellent knowledge of FIX and messaging based connectivity applications. Currently in last year of PhD in Virtual Robotics at the University of Krakow. Three years of work experience at Lorien Engineering Polska.

Regards

Simon

ABC Recruitment Ltd

All Simon has done is to paste Juri's executive summary (the part in italics) into the email. This saves Simon a lot of time.

So by including an Executive Summary in your CV, you may increase your chances of a recruiting agency sending your curriculum to a firm.

6.11 I am in research. Do I really need an Executive Summary?

At the time of writing this book, executive summaries are not commonly used in academia, but would in my opinion be equally useful.

An alternative for academics is to put at the top of your CV, your name followed by your position. Here are some examples:

Sanchez Panza
Professor of Genetics

-

Ekaterina Milovski
Full-time researcher in animal psychology

-

Hao Pi
PhD student in Business Studies

Or you could write an objective. So Ekaterina Milovski could write:

A researcher in animal psychology seeking a permanent position in a university veterinary hospital.

Hao Pi could write

A PhD student in Business Studies looking for a six-month internship in a commercial bank.

6.12 What is a Personal Statement? What are the elements of a good Personal Statement?

A personal statement is an optional section, typically written by candidates who have finished their education with a normal degree without doing an MSc, a PhD or a post-doc qualification. Personal statements are also used by school leavers when applying to university.

The aim of a personal statement is to show that you:

- have the right qualifications for the place / position you are seeking

- have the right skills - both technical and personal (i.e. soft skills)

- can describe yourself and your achievements concisely.

6.13 What are the typical downfalls of a Personal Statement? What should I avoid?

Below is a personal statement written by a young British graduate who wishes to secure a job in teaching English as a foreign language (TEFL) in a language school. It highlights some of the good and bad points of a typical personal statement.

Personal Statement:

Having a BSc degree has given me the skills required to use language to a high standard and write and communicate with many different people. Among the modules I studied were:

- Environmental Management

- Environmental Law

- Information Technology and Quantitative Biology

- Environmental Economics

I have worked primarily in the customer services area either in shops or on a campsite, which has given me good experience of working with people in teams and offering an appropriate quality of service to customers. I have completed a TEFL course in order to teach English; completing this course has helped my skills in planning, listening, speaking to groups, and ensuring understanding. Since then I have undertaken more teaching qualifications and employment all of which have been very successful; I have been well received by students, fellow teachers and course organisers. Recently I have been more and more involved in TEFL teaching having been working both in my home city of Bristol and in Prague. I would really like to extend these experiences to working abroad, especially in your city, to which I have a particular attachment having enjoyed my stay previously.

I am committed to finding for myself a career, but I need an opportunity. I hope that you will give me an opportunity to start a career with you and within the industry.

6.13 What are the typical downfalls of a Personal Statement? What should I avoid? (cont.)

Her layout is clear using bullet points, she clearly lists the main topics she studied at university and she tries to connect this to TEFL by saying that she has strong skills in writing and communicating. However there are a number of problems:

- the statement is very long, although she claims to have good writing and communication skills, her statement shows that she is prone to repetition and redundancy. She says she has completed a TEFL course and then adds *in order to teach English* - this is redundant given the fact that the TE in TEFL stands for 'teaching English', the same redundancy is in 'TEFL teaching'. However, this tactic of repeating her key words (i.e. TEFL, teaching and English), means that if she submitted an electronic version of her CV to a recruiting agency website, it might stand a better chance of being picked up during any automatic screening process of CVs

- the statement does not seem to have been tailored specifically for the reader - she says *especially in your city*. If she is writing directly to a language school, she would be better to write the exact location of the language school and mention the exact time she visited that town, rather than writing something so generic that gives the idea that she has probably never even been to *your city*. Moreover, the fact that she used *your city* shows that she hasn't really made an effort, and this could be interpreted by the reader that she would be happy to work in any city

- in her final paragraph, she sounds a little desperate (*I need an opportunity*) and also the phrase *I am committed to finding for myself a caree*r sounds a little strange, given that securing a career is the objective of most graduates. Again it sounds like she may be applying for jobs in any field, not the specific field of TEFL

Below is a revised version. The main differences with respect to the original version are highlighted in italics.

Personal Statement:

My BSc has provided me with the skills required to use language to a high standard and write and communicate with many different people. Among the modules I studied were:

- Environmental Management

- Environmental Law

- Information Technology and Quantitative Biology

- Environmental Economics

6.13 What are the typical downfalls of a Personal Statement? What should I avoid? (cont.)

I have worked primarily in the customer services area both in *shops and campsites*. This has given me valuable experience of working with people in teams and offering an appropriate quality of service to customers.

During my TEFL course I further improved my skills in planning, listening, speaking to groups, and ensuring understanding. Since then I have *acquired* more teaching qualifications and *have successfully gained additional teaching experience in my home town of Bristol and in Prague. As testified by the attached references,* I have been well received by students, fellow teachers and course organisers.

I would really like to extend these experiences to working abroad, especially in Moscow, to which I have a particular attachment - *my maternal grandmother was born in a village near Moscow.*

I very much look forward to having an opportunity to meet you.

The revised version:

- is 20 % shorter. All the repetition and redundancy has been removed, but no content has been lost

- has more paragraphs - this makes it easier to read

- has removed ambiguity (e.g. *acquired* teaching qualifications, rather than *undertaken* - *undertaken* sounds like the courses were started but not completed)

- mentions *references* to give her more credibility, i.e. what she says can be supported by the people she has worked for

- mentions the specific city where she wants to work (Moscow) rather than saying *your city* (this generic use of *your city* could indicate to the recruiter that the candidate has sent the same personal statement to lots of English language schools. A tailored personal statement has more impact

- has deleted the rather strange final paragraph

The combination of the above changes makes the statement sound as if the candidate had put a lot of effort into creating it. It also makes the candidate sound more dynamic.

6.14   Any differences in a resume?

A typical resume contains a short summary at the top of the resume (i.e. the first section after the candidate's name). This summary is variously entitled: Objective, Personal Objective, Executive Summary, Career Highlights, Career History, Qualifications Summary.

The information contained in this section (whatever its name) is identical to what you would write in the same section in a CV. You can also use any of the suggested formats mentioned in 6.6.

Summary: Objectives, Executive Summaries, and Personal Statements

Objective

➢ One sentence summarizing what kind of job you are looking for (so generally not for a position that has been advertised).

➢ Ensure that it is not ambiguous and does not focus exclusively on the benefits for you.

Executive summary

➢ A summary of the work, education and skills sections of your CV. Typically in response to an advertised position.

➢ Try to create enough space to use a series of bullet points each of which matches the order of the job spec.

➢ If you are not responding to a specific advertisement, then you can save space by using a single paragraph.

Personal statement

➢ Similar to an executive summary, but typically used by university applicants or first time job hunters.

In all three cases:

➢ Insert key words (i.e. those from advertised job spec, or from job specs in your field).

➢ Tailor it to your chosen organization.

➢ Be concise.

7 EDUCATION

7.1 Where should the Education section be located and what should it include?

If you finished your education several years ago, this section should appear after your Work Experience section and should contain fewer details than in the Work Experience section.

The section on education should include the:

- start and end dates (see Chapter 3); everything should be in reverse chronological order

- name and location of the institute, plus a web link to the institute / department

- type of degree

- brief details of coursework

If you are sending your CV outside your own country, then the reader may not be familiar with the level of prestige of the university / organization you attended. So provide a link to a relevant page in the university's website (see an example for Brunel University in 7.2).

A. Wallwork, *CVs, Resumes, and LinkedIn*,
Guides to Professional English, DOI 10.1007/978-1-4939-0647-5_7,
© Springer Science+Business Media New York 2014

7.2   What is the typical layout?

Below is a typical layout.

	Education and training
2028-2031	PhD (ISCED 6), funded by an Economic and Social Research Council Award at Brunel University, London, UK (www.brunel.ac.uk/) Thesis Title: 'Young People in the Construction of the Virtual University', empirical research that directly contributes to debates on e-learning.
2024-2027	Bachelor of Science in Sociology and Psychology (ISCED 5), Brunel University, London, UK (www.brunel.ac.uk/) Principal subjects/occupational skills covered: Sociology of Risk, Sociology of Scientific Knowledge/ Information Society; E-learning and Psychology; Research Methods.

Note how the candidate has given a link to his university.

Links are useful because they enable the reader to learn more about the level of prestige of the university / organization you attended. Instead of a link to your university's home page (or the company's homepage in your Work Experience section), you can provide a link to your department, or your research / work group.

You may also consider adding some extra information about your university, e.g. where it appears in the world's academic ranking.

Another link you could include in this section is to the full version of your thesis.

7.2   What is the typical layout? (cont.)

Here is another possible layout.

Education

2020– 2023 University of Manchester, UK

Doctor of Philosophy in Information Engineering

Research in greening the Internet. Elective coursework included: enhanced Internet architecture employing advanced communication service paradigms, protocols and algorithms targeted at the optimization of energy consumption. Dissertation "A radical energy-aware application for wireless energy reduction" advised by Professor Giuseppe Verdi.

2017-2020 University of Santiago, Chile

Bachelor of Science degree (5 year course) in Electrical Engineering

Engineering coursework included: continuous and discrete systems and signal processing, analog and digital circuit design, and computational theory. Undergraduate thesis project, "Wireless Enabled Context Awareness for the Future Internet".

Note from the example above that:

- the information is in reverse chronological order – this is mandatory in a CV

- the candidate has not mentioned what school she went to (see 7.3), instead she has begun directly with her university education

- the candidate got her BSc in Bolivia and has also indicated how long the course was (5 years) – this is important as it shows that her course was longer than the standard 3-year course

- she has just written her thesis title, without describing any details – this is because in both cases her thesis title is self-explanatory

7.3   Do I need to mention my high school?

If you finished your education after your first degree and are now looking for your first job, then putting your high school is perfectly normal and acceptable. However, if you subsequently did a Master's or PhD then mentioning your school is probably not necessary.

The name of your school is very likely to be in your native language and will thus give no useful information to the hirer. In such cases you can do as follows:

2017-2021 Secondary school specialized in scientific studies. Final score: 18/20

Here are some more examples of possible specializations at secondary school:

specialized in *the classics* or *classical studies*

specialized in *business* or *business studies*

specialized in *languages* or *linguistic studies*

specialized in *literature*

If your secondary school had no specialization then simply write:

2017-2021 Secondary school – generic studies. Final score: 18/20

If your final score is expressed as a percentage, then you can write:

2017-2021 Secondary school specialized in business studies. Final score: 88%

7.4 I am not sure whether my degree has an equivalent outside my own country. What should I do?

Every country has its own system of naming and classifying degrees. On Wikipedia there are full descriptions of the types of degree (Bachelor's, Master's, PhD) you can obtain in the UK and the USA.

The equivalents between first degrees in the UK and the USA are listed in table below, which comes from the University College London website (www.ucl.ac.uk/).

UK	USA / Canada
First-class Honours	GPA (grade point average) 3.6/4.0
Upper second-class Honours	GPA 3.3/4.0
Lower second-class Honours	GPA 3.0/4.0

When you list your academic qualifications it may be worth trying to explain what your degree is equivalent to in the country where you are sending your CV. For example, imagine you have done your first degree in Mexico and you have decided to send your CV to two institutes: one in the USA and one in Germany. You can look on Wikipedia to see what is the closest equivalent to your Mexican degree, and you will discover that it is called a Bachelor's (http://en.wikipedia.org/wiki/Bachelor's_degree). You then look for descriptions of Bachelor's degrees in the USA and Germany. Obviously, you will write two CVs, one for each country. On the one for the USA, you can describe your degree by using a direct equivalent for the score you got.

2024-2029	Bachelor of Science in Telecommunications, Universidad Nacional Autónoma de México- score: 98/100 cum laude (equivalent to Bachelor in Germany, note: the duration of the

To prepare the CV for Germany you can look at Wikipedia's entry for Germany (http://en.wikipedia.org/wiki/Bachelor's_degree#Germany) where you will discover that the name of your degree in German is either *Bakkalaureus* or *Bachelor* and that in Germany this requires three years of study. But maybe your degree lasted five years, so in this case you need to state the difference in length.

2010-2011	Master's in Particle Physics (full description at http://english.pku.edu.cn/blahblah), University of Bejing, China

7.4 I am not sure whether my degree has an equivalent outside my own country. What should I do? (cont.)

If you are unable to find an equivalent to your degree, then you could also put a link next to your qualification to a website where there is a full description of your degree in English. Note: in the example below the website is fictional.

<u>VGB Volunteers</u>	In order to obtain some teaching experience, I am going to undertake a volunteer
English Teacher	placement with a charity in Surin, Thailand for a month. The charity offers free
July - August	English lessons to tuk-tuk drivers and civil servants, for whom the expansion of
Surin, Thailand	their knowledge of English will enhance their employability greatly, given the
	importance of tourism in Thailand.

The Bologna Process aims to have a common description of degrees/standards inside Europe, see http://en.wikipedia.org/wiki/Bologna_Process.

Various other websites also list degree equivalents between the UK and the rest of the world: http://www.how2uk.com/ukstudy/t/266/equivalent-degree-classification-for-uk-unis-and-your-country/. You can the use this data to make comparisons between your degree and a US equivalent or an equivalent in another country.

7.5   How should I write my score?

How you write you score involves a similar problem as to how you describe your degree. The problem is that the score can easily be interpreted by someone who is familiar with your system (i.e. because they are the same nationality as you) but to someone from another country it may be meaningless.

For example, is a score of 89/100 a good score? It seems high, 89%. But in Italy, where this scoring system exists, 89/100 would not be considered a particularly good score. A top student would get 100/100, and the absolute best would be '100/100 cum laude'. In the UK a score of 100/100 seems very unlikely, particularly in a non-scientific subject where there are no exact answers.

There are two possible solutions.

1. Do not put the score, but put the equivalent qualification for the country where you are seeking employment (see the table in 7.4).

2. Put the score, but interpret it for your readers. For example, if scores in your country are recorded as a percentage then you could write that you obtained a score of 89% and that a score of between 85%-90% is only achieved by one in twenty students. This then gives the reader an idea of how good you are.

7.6   How much detail should I give about my thesis / dissertation?

What you write with regard to your thesis will depend on i) how recently you defended your thesis, and ii) how much space on your CV you have available.

Only put your thesis title if it is self-explanatory. If it is not self explanatory, you can write, for example:

> Thesis on new methodologies for extracting gold from recycled plastic. The procedure involved three steps ...

In any case the description of your thesis should not be more than two lines long.

There is no need to write the exact date you defended you thesis (unless this is very recent). Unless you want to fill up space on your CV you don't need to mention the names of your supervisors or the committee chair and members. However, if these people are known personally to the institute / company where you are sending your CV, or are very famous in their field, then it is certainly worth mentioning them. In any case, you can give details of these people by providing links to their web pages.

7.7   What about additional courses that I have attended?

In addition to your main degree courses, you may wish to include other courses that you attended that are pertinent to the job you are applying for and which would be beneficial to the position you are seeking.

Below is an example from a recent graduate:

Manchester Academy of English – CELTA (Pass)

The course focused on: x, y, z. It highlighted the importance of seeing things from the learner's perspective in order to gain insights into how best to teach English. We undertook six hours of assessed teaching practice, observed six hours' worth of experienced teachers' lessons, and completed four assignments. Studying for the course part-time whilst holding down two jobs vastly improved my time-management skills and taught me how to manage my workload effectively.

Note how he does not merely say what the course consisted of (*x, y and z*) but also the benefit for the candidate (*importance of seeing …*). He takes the opportunity to show how the fact that he did the course and worked at the same time highlights that he now has particular skills (*time management*) that will be relevant for the job he is applying for.

He uses a personal style throughout. If you choose to adopt such a style, then be very careful as you are more likely to make errors in English.

7.8   I am a recent graduate. My CV looks rather empty. What can I do to fill it up?

If you are student and your CV looks a little empty, then you might also consider including a future experience that you have arranged for the very near future. The example is of a student from England who wrote his CV in May and was hoping to get a job as an English teacher in October of the same year:

VGB Volunteers	In order to obtain some teaching experience, I am going to undertake a volunteer
English Teacher	placement with a charity in Surin, Thailand for a month. The charity offers free
July - August	English lessons to tuk-tuk drivers and civil servants, for whom the expansion of
Surin, Thailand	their knowledge of English will enhance their employability greatly, given the
	importance of tourism in Thailand.

7.9 Is it worth mentioning my teaching experience, even if it does not directly relate to the post I am applying for?

Yes, definitely. If you have taught some undergraduate classes, you will have learned some useful skills while in the classroom, for example how to:

- stand up in front of an audience and overcome your nerves

- explain difficult concepts in a simple way

- manage people

- prepare lessons and presentation slides

- work within specific time frames

These are all very useful and transferable skills that any employer either in industry or academia will appreciate.

Below are two examples. Note that in both cases impersonal forms have been used. However, the candidate could also have written: *I composed, I advised.* One advantage of the impersonal style is that it saves space.

Teaching Experience

MIT Department of Physics Cambridge, MA

September 2025 – December 2025

Teaching Assistant

Taught sophomore level course on Quantum Mechanics.

Advised Anna Southern on her MIT undergraduate Physics thesis. Anna developed and applied experiments for identifying ...

Teaching Experience

September 2025 – December 2025, Teaching Assistant, MIT Department of Physics Cambridge, MA

- Sophomore level course on Quantum Mechanics.

- Advice to MIT Physics undergraduate for thesis.

7.10 Any differences in a resume?

Yes, there are differences. This section will typically be found at the end of the resume, or in penultimate position if the resume also contains a Personal Interests section.

The layout should be the same as in the third layout outlined in 7.2. If you already have considerable work experience, then in this section you would probably just list the name of the institution and the qualification you received, without any further details.

Summary: Education

➢ Put each experience into a separate mini section.

➢ List as follows: i) date ii) place iii) qualification (i.e. type of degree, certificate) iv) details of qualifications - provide links to relevant webpages so that HR can understood more about the university / qualification

➢ Dates: reverse chronological order.

➢ High school: only mention if you are a recent graduate.

➢ Scores and degrees: if the scoring / degree system of your country will not be immediately clear to the reader, interpret your score and / or provide an equivalent for the country where you are applying for a job.

➢ Thesis: maximum two line description.

➢ If your CV looks empty: mention additional courses, teaching experience, future plans.

8 WORK EXPERIENCE

8.1 Where should the Work Experience section be located? What's the best layout?

If you finished your education several years ago, this section should appear before your Education section and should contain more detail.

Note that in this section you should also provide some evidence of your soft skills – see 9.7 to learn how to integrate such skills into your Work Experience section.

You can describe your work experience (i.e. within a company) in exactly the same way as for the Education section (see Chapter 7). Simply put the name of the company, its location and your position within the company. Then write a description of what you did there, highlighting how it fits the requirements of the job you are applying for.

As in the Education section, everything should be in reverse chronological order.

2029 – now.
University Lecturer – Tshwane University of Technology (www.tut.ac.za), South Africa

- Preparing and teaching software engineering subjects, such as Systems Analysis and Design, OO Analysis and Design using UML, SQL & PL\SQL, Data Engineering, OO Programming, Object-Oriented Software Engineering (Software Development as an engineering field), Project Management.

- Designing and developing distributed Systems (for 2nd and 3rd year students) and Advanced PL\SQL for Final Year (4th) Computer Science and Software Development Students.

2027–2029
Information Systems Specialist (Intern) – Vodacom Congo SPRL, DRC

- Wrote and debugged programs and complex SQL queries for customer consumption forecasting.

A. Wallwork, *CVs, Resumes, and LinkedIn,*
Guides to Professional English, DOI 10.1007/978-1-4939-0647-5_8,
© Springer Science+Business Media New York 2014

8.1 Where should the Work Experience section be located? What's the best layout? (cont.)

- Produced regular statistical reports about the activities of subscribers with regards to calls made, credit recharging, and all SIM Card related activities.

- Developed a system to help manage the distribution and coordination of tasks between the company and its different dealers, retailers.

Note how the candidate has provided a link to his current workplace. This enables the HR person or recruiter to learn more about the workplace, without the candidate having to waste a lot of space giving explanations.

The above candidate is a PhD student, yet he has managed to include references to work experiences. You can include the following under 'Work Experience':

- internships

- lecturing / teaching

- summer jobs

Highlight that one experience was more important than another by allocating it more lines. For example, given that an internship is probably more relevant than a summer job, it should have more space dedicated to it. This is a general rule for your whole CV: the more important the experience (from an HR perspective), the more space should be dedicated to it relative to less important experiences.

Below is another good example. In this case, the candidate below has organized her Work Experience as follows:

- first line: dates + name of company (plus a web link to the company, so the HR person can find out more about where you work / have worked)

- second line: her position within the company

- bullet points: each indicating key roles that she carried out

The amount of detail depends on how much space you have available.

8.1 Where should the Work Experience section be located? What's the best layout? (cont.)

2026 – present: Zed Engineering Group (zedenggroup.com), Kuala Lumpur

Principal Engineer – Secure Systems

- Secured $500,000 of funding for research into MegaData stream processing architectures. As part of this project, I developed a Java stream processing application running on the S5 platform for detecting algorithmically generated domain names in DNS queries. I also standardized procedures regarding 'Scalable and Elastic Event Processing' (SEEP) and secured applications by enforcing Information Flow Control policies within middleware.

- Architected and developed innovative security architecture, written in core Java, for a test-bed sharing coalition Intelligence, Surveillance and Reconnaissance data at different classifications.

2020 – 2026: Tata Engineering, automotive department

Software Architect

- Controlled the software architecture for real-time embedded systems in the Driver Information product line.

- Reviewed work products: implementing critical features (using object-oriented C and Java), specifying the development methodology and processes, and troubleshooting problems.

- Supported new business wins with major car manufacturers.

8.2 Which is better *I developed a system* or *Developed a system* (i.e. with or without the personal pronoun)?

As highlighted in the example in the previous subsection, you can use verbs without a subject at the beginning of each bullet point (e.g. *secured, architected, developed, controlled*). Instead, if the verb does not come at the beginning of the bullet point, then you would probably use the personal pronoun before the noun (highlighted in italics in the example below):

> As part of this project, *I developed* a Java stream processing application running on the S5 platform for detecting algorithmically generated domain names in DNS queries. *I also standardized* procedures regarding 'Scalable and Elastic Event Processing' (SEEP) and <u>secured</u> applications by enforcing Information Flow Control policies within middleware.

Note that the verb underlined (*secured*) has no subject as it is implicit because it is within the same sentence as *I also standardized*.

8.3 How can I highlight how my work experience fits in with the post I am applying for? What key words should I try to insert?

Most companies and recruiters use applicant-tracking systems in order to scan CVs for key words. The key words that the systems are searching for will be the same key words that appear in the job description.

So if you are applying for a job which has been advertised, analyse the job description and decide what the key words are. Then try and insert these key words in the most natural way possible into your CV (see 6.4, 6.8, 8.3, 8.4 and 14.5).

Basically, the more matches the system finds between the job description and your CV, the more likely your CV will be read by a real person.

8.4 How can I make my key words stand out, yet not be too obtrusive?

Let's imagine that you want to further your career in the field of web management. Your objective in your CV (and in your LinkedIn profile, see 14.5) is to fill your CV with keywords connected with web management, but without making it too obvious. Here is a good example of how to achieve that aim.

Notice how the candidate writes some of his key words with initial capital letters (e.g. Web Producer). This makes his key words stand out to the human reader (but obviously makes no difference if his CV is simply being scanned automatically).

Head of Digital: July 2029 – present

Marketing Media plc, 15–25 Newton Street, Manchester, UK, M1 1HL

Leading a team comprising a Web Producer, Web Developer and a Film Producer/Editor, responsible for the planning, strategy and delivery of a careers advice and inspiration website.

- Developed e-communications strategy to grow the Creative Choices audience
- Successful negotiation with Google to sponsor a Google Adwords account
- Set and monitored quality, accuracy and style guidelines for web content
- Developed new digital tools and web services, from idea stage, through specification to delivery

Web Manager 2024 – 2029

XYZ Legal Consulting, 17 Whitley Road, Newcastle Upon Tyne, UK. NE98 1BA

Developed and maintained online content; responsible for:

- research, design and compilation of legal compliance information delivered online
- the style, accessibility and accuracy of all content on the web.

How many times do you think the word *web* appeared in the above extract? Three, four, five times? The candidate has quite subtly managed to include his key word seven times (including *website*) – to the human eye it won't be too intrusive, and at the same time a recruiter's software will be able to find multiple instances thus increasing the candidate's chances of having his CV selected.

8.5 I am a recent graduate. How can I describe my work experience in shops, restaurants, hotels etc. in a very constructive light?

As always, the important thing is to be totally honest (see 1.11). However, you are also trying to 'sell' yourself to recruiters, so you can make your work experience sound relevant and worthwhile by adding relevant details.

Below is an extract from the Work Experience section of a CV of a 23 year-old girl who went to London to learn English and gain some work experience. She worked as a sales assistant in several clothes shops, and each time she changed job she managed to secure a slightly better position and salary. In her CV she manages to make the most of what was in reality fairly routine work, but which nevertheless contributed to sound work experience. She also manages to highlight the various sales and communication skills that she acquired.

Note how she devotes more space to her most recent experience, which thus gives a sense of her 'career' progression.

October 2029 – present, Aubin & Wills, Selfridges, Oxford Street

Position: personal stylist. Main activities: styling and fitting service. Developed knowledge of vintage clothing and British textile manufacturers. Looking after customers; offering advice and styling according to customer needs and requests. Excellent customer service and advanced cashier operations (dealing with customer complaints, exchanges, refunds, cashing up). Participated in live music nights in store, magazine and bloggers' events for the company's product launches as a denim consultant.

March 2029 – October 2029, Nigel Hall Menswear, Selfridges, Oxford Street

Position: personal stylist. Main activities: styling and fitting service. Looked after customers, offering advice and styling according to customer needs and requests. Developed knowledge of formal wear and styling. Excellent personal target achievement and customer service.

March 2028 – March 2029, London Levi's Flagship Store, Regent Street

Position: denim expert and fitting consultant. Main activities: knowledge regarding the history and the treatment of denim, vintage clothing and different kinds of styling. Sewing and customizing skills (alteration service). Took part in several visual merchandising projects in store.

8.6 How should I describe internships and other research experiences?

When you describe what you did during your periods of work / research experience, ensure that what you write provides evidence that you have relevant experience for the job you are applying for.

Below are two examples of how to describe an internship, i.e. a period of supervised training at a research laboratory or in a company.

EXAMPLE 1

	Internships and summer schools
Apr - Dec 2024	Internship at the European Space Agency, Paris led by Dr Spock. My research was part of the standardization process of DVB-S2. I studied both theoretical and computer simulation of the carrier phase and frequency recovery schemes for DVB-S2 applications. The challenge was to explore low complex synchronization solutions for dealing with the high level of transmitter/receiver oscillator's phase noise. The outcome of this work was submitted for several ESA patent applications. I managed a small group of researchers for part of the internship.
Jun - Aug 2023	Summer school at NASA, Houston, Texas. I worked alongside several scientists and astronauts and was trained in the following areas:

In Example 1, the candidate uses personal forms and verbs: *my research, I studied* and *I worked*. On the other hand, in Example 2 below the candidate uses a series of nouns (*analysis, development, implementation*) with few verbs. Both these styles are typical, the second is less likely to lead to mistakes in grammar.

8.6 How should I describe internships and other research experiences? (cont.)

Research **Experience**	HITECH HYTO LABORATORY, KYOTO, JAPAN
	2023 - present
	Postdoctoral scholar, currently Visitor in Physics
	Gravitational-wave data analysis. Development and implementation of algorithms to search for unmodeled bursts of gravitational radiation in data from interferometric detectors. Member of the HYTO Scientific Collaboration (HSC). Co-chair of HSC glitch working group and member of the PHYTO analysis group.
	QTX CENTER FOR SPACE RESEARCH, OXFORD, UK
	July 2022 - Jan 2023
	Sponsored Research Technical Staff
	Design, construction, and operation of an experimental apparatus to measure correlated magnetic and seismic fluctuations between the two QTX observatory sites.

8.7   I have done some jobs that don't seem to fit under the heading Work Experience, can I call them 'Other Work Experience'?

Ideally, you don't want too many headings in your CV. So try and fit all your work experiences under one simple heading.

However, if you have little work experience and your CV looks rather empty, then you might want to fill it up with a new section. Here is an example:

OTHER WORK EXPERIENCE

Aug 2030: Research Internship at the New Policy Institute, London

- During this one month internship at a think-tank in London I helped to compile a major report about poverty and social exclusion within the UK. In particular I was investigating the effect of poverty and inequality on crime rates and health.

Nov 2028 – May 2030: teaching assistant, Altrincham Grammar School, UK

- I worked once a week at a grammar school in Cheshire with a group of five students aged 15–17. My role was to act as an informed and resourceful role model and to give the students a different perspective on their studies. This included helping the older students with career advice and university applications, and showing the younger students a range of options of higher education such as applying for college and apprenticeships.

- I worked both one-to-one and with groups of students and I taught them how to use various online resources which would help them with their studies.

Some jobs that you have done (or still do) may have nothing to do with your career path. You may have done them simply to earn money during the holidays or to support your expenses at university. Such jobs include babysitting, working in bars and shops, and doing voluntary work. They are worth mentioning because they indicate that you are:

- a responsible person (if some parents leave their child with you they must consider you to be responsible and mature)

- able to wok with all kinds of people

- independent – you don't just rely on your parents to give you money

Rather than having a separate section, you can list these jobs under Personal Interests (Chapter 10).

8.8 I am applying for a job in industry, do I need to have a list of my publications?

You don't need to have a list of publications or a separate section dedicated to publications – this is only needed if you are applying for a job in research/academia.

However, the fact that you have published your research is still important, even to an employer in industry. Publishing your work means that you have certain skills:

- writing in English about technical matters
- communicating with referees and editors, so you will have written many formal emails and letters
- meeting deadlines
- presentating your paper/research at international conferences

So the solution is to add a short subsection to your Education or Work Experience section in which you write something like this:

> First author of five papers on civil engineering, published in international journals. Presented three of these papers at international conferences. Papers available at: www.blahblah/blah

8.9 I am a researcher. Where should I locate my publications?

If you have a short list of publications (four or five works) you can simply put them under one section entitled 'Publications' and locate it directly under either 'Other skills' or 'Hobbies and interests'. You will then follow 'Publications' with your 'References' section (Chapter 11) which will conclude your CV.

If you have a long list of publications, it is probably best to start the 'Publications' on a separate page. This then helps the main part of your CV to stand out and not to go over two/three pages (the recommended number of pages for a CV).

8.10 Do I need to divide up my publications into various subsections?

You might also consider dividing up your publications into the following subsections:

- *Selected Refereed Publications* – these are the ones you want the reader to focus on

- *Other Refereed Publications* – these extra ones help to highlight the quantity of research that you have had published

- *Pending Publications* – these are ones that either you have submitted (and are awaiting confirmation) or that are currently at the press

- *Technical Notes* – these are short articles outlining a specific development/modification, technique or procedure

You should list your publications in the same way as you would normally list the publications at the end of a paper. With regard to pending publications you can write:

A. Wallwork et al. "Detailed comparison of word order in Modern and Old English". To appear in Annals of Ling. Rev.

A. Wallwork et al. "The subjunctive in Old English texts". Submitted to Int. Lang. Rev.

The term to *appear* in means that your paper has already been accepted for publication, whereas *submitted to* means you are waiting for the outcome.

8.11 Any differences in a resume?

Yes. This section will typically be found immediately after the Objective / Executive Summary, i.e. before your Education section.

The layout should be the same as in 8.4 and 8.5.

The only main difference in style is that resumes tend to avoid the use of personal pronouns (highlighted in italics below). So instead of saying (see 8.1):

> As part of this project, I developed a Java stream processing application running on the S5 platform for detecting algorithmically generated domain names in DNS queries. I also standardized procedures regarding 'Scalable and Elastic Event Processing' (SEEP) and ...

You would simply avoid the use of the first person pronoun *I* and write:

> As part of this project, was responsible for the development of a Java stream processing application running on the S5 platform for detecting algorithmically generated domain names in DNS queries. Also, was in charge of setting up standardized procedures regarding 'Scalable and Elastic Event Processing' (SEEP) and ...

Summary: Work Experience

➢ Separate each experience.

➢ List as follows: i) date ii) company / organization iii) position iv) key roles played (plus a web link to the company, so the HR person can find out more about where you work / have worked)

➢ Dates: reverse chronological order.

➢ Grammar and conciseness: verbs without pronouns at beginning of sentences (e.g. *Developed* rather than *I developed*).

➢ Insert as many key words as is reasonably possible.

➢ Key words should be those in the advertisement for the job you are seeking, or on similar job specifications for people in your field if you are not applying for a specific job.

➢ Do not use personal pronouns in a resume.

➢ Make your experience sound both relevant and dynamic.

➢ Only have a separate publications section if you are applying for a job in academia. Consider having this section as a separate document (i.e. not a direct part of your CV) or simply use a link to a webpage where your publications are listed.

9 SKILLS

9.1 Should I mention all my technical skills?

To answer that question, look at the extract from a CV below. What impression do you have?

Software Skills	Good knowledge of Matlab / Simulink.
	Good knowledge of C/C++ language.
	Good knowledge of Java language.
	Good knowledge of Html/Javascript.
	Good knowledge of ASP.
	Good knowledge of PHP.
	Good knowledge of Visual Studio.
	Good knowledge of query language (MySql).
	Good knowledge of Unix operating systems (including Freebsd, Kubuntu and Debian).
	Good knowledge of Latex.
	Good knowledge of Doxygen.
	Intermediate knowledge of Labview.

There are many problems with the example above. The most obvious are the use of a horizontal rather than a vertical list and the repetition of 'good knowledge'. It would be much more concise to do as follows:

Software Skills

I have good knowledge of the following: Matlab / Simulink; C/C++; Java; HTML/Javascript, ASP, PHP

Intermediate knowledge: Labview.

The other problem is credibility – is it possible to have 'good knowledge' of so many systems and languages? Just list the technical skills that are listed in the company's job specifications or which you think might in any case be useful for the job you are applying for.

A. Wallwork, *CVs, Resumes, and LinkedIn,*
Guides to Professional English, DOI 10.1007/978-1-4939-0647-5_9,
© Springer Science+Business Media New York 2014

9.2 Under what section should I put my language skills? And how do I mention them?

You can either have Language Skills as a separate section. Or you can have a section entitled Skills under which you put your language skills and technical skills.

If you are applying for a position in Europe, you can use the levels from the Common European Framework of Reference for Languages. Below is an example from the 'Personal skills and competences' section of the Europass CV.

Mother tongue(s)	**Norwegian**								
Other language(s)									
Self-assessment		**Understanding**				**Speaking**			**Writing**
European level (*)		Listening		Reading		Spoken interaction		Spoken production	
English	C1	Proficient user	C1	Proficient user	C2	Proficient user	C2	Proficient user	B1	Independent User
French	A2	Basic user	B2	Independent user	A2	Basic user	A2	Basic user	A1	Basic user

For more details on how levels of language proficiency are classified see: http://en.wikipedia.org/wiki/Common_European_Framework_of_Reference_for_Languages

Given that the Europass format takes up a lot of space, as an alternative you could use these terms:

- mother tongue
- fluent (spoken and written)
- good working knowledge – means that you know enough to be able to carry out your work
- scholastic (typical level achieved at school – but only mention such languages if they are specifically mentioned in the job specifications and you feel that they would increase your chances of getting the job)

9.2 Under what section should I put my language skills? And how do I mention them? (cont.)

Thus, for example, you could write:

Languages	
Korean	Mother tongue
Chinese	Fluent
English	Spoken (good working knowledge), Listening (independent user), Written (proficient), Reading (proficient)

If you are not using the Europass, then you could write:

Korean: mother tongue; **Chinese**: fluent; **English**: spoken (good working knowledge), listening (independent user), written and reading (proficient)

If you are very short of space, you can list the languages you know in your Personal Details section. For example:

Kamran Kamatchi

914 West 10th Street, Hazleton, PA, 18209, USA

k.kamatchi@gmail.com (+11) 7525 446779

Languages: English (native), Bengali (native), Spanish (good working knowledge)

9.3 What about English examinations I have taken?

One way to prove that you have good language skills is to list the English examinations that you have taken, for example the TOEFL exam (www.ets. org/toefl) and the Cambridge exams (www.cambridgeenglish.org/exams-and-qualifications/).

The TOEFL exam is for American English, and the Cambridge for British English. They have very different formats, and the Cambridge exams have many different levels. The TOEFL certificate only has a limited duration, so in theory you are supposed to keep retaking it. The Cambridge exams have unlimited validity.

For the purposes of your CV you can simply put the exam and the grade, plus a link to the relevant website to allow the HR person to see exactly what the examinations entailed.

If you have done more than one level of the Cambridge exams, only put the highest level. For example if you have done the First Certificate (B2, intermediate level) and the PET (B1, low level), only list the First Certificate.

Below is an example:

Languages	
Russian	Mother tongue
Latvian	Mother Tongue
English	Cambridge First Certificate (Grade A): www.cambridgeenglish.org/exams-and-qualifications/first/

9.4 I passed an English examination years ago, and since then my English has probably got worse. Should I still mention the exam?

Mention the exam, but not necessarily the year.

Before the interview, try to get your English back to the level of the examination.

9.5 Are there any advantages of mentioning languages for which I only have a basic knowledge?

Only for two reasons:

• if the job specification specifically states that knowledge of that language would be 'useful'

• if your CV is relatively empty and you need to fill up the space

Otherwise, don't mention them.

9.6 What are the risks of exaggerating my language skills?

High.

If you claim that you have an 'advanced' knowledge of English but there are mistakes in the English of your CV and cover letter, then you will immediately lose credibility. Also, if all or part of the interview is conducted in English (either over the phone or face to face) and you fail to perform to an advanced level, the interviewer may suspect that you may have exaggerated other parts of your CV too.

So be honest.

If you are not sure of your level, then get a qualified EFL / ESL teacher (English as a Foreign Language / English as a Second Language) to give you a quick test.

9.7 Should I have a separate section entitled 'Communication Skills'?

No.

You may have acquired many communication skills and not necessarily directly during research or employment. In any case such skills will probably be useful for the position you are seeking. However these skills should be implicit in the Education, Work Experience (see end of this subsection) and Personal Interests (10.4) sections of your CV as well as in the cover letter and reference letters (see 11.16). They should not be listed in a separate section.

Below is an extract from a section entitled 'Personal Skills and Competences'. What would the HR person learn from the information given? Basically, that this candidate is the same as every other candidate. The problem is that lists of such skills are completely pointless unless substantiated by evidence.

Social skills and competences	Ability to build relationships easily I solve problems easily and efficiently. Team player
Organisational skills and competences	Experience in organizing groups of students
Other skills and competences	Ability to work under pressure

Below is an example of how to integrate some of your soft skills into the body of your CV, in this case the Work Experience section. For the purposes of this book, the soft skills are highlighted in italics.

Jul 2027–Sep 2028: Fermilab – Fermi National Accelerator Laboratory, Batavia, IL, USA

Worked as a Computer Engineer at Fermilab in the Accelerator Division under the supervision of Brian Chase and Paul Joiremann. *Headed up various small internal work groups.*

Fermilab conducts basic research into particle physics. *Part of my duties included technical presentations of Fermilab research projects to interested partners.*

Development of hardware-software interfaces and client-server interfaces using C / C++, DOOCS, Labview. *This included solving critical problems in very short timeframes relating to high-energy physics.*

9.7 Should I have a separate section entitled 'Communication Skills'? (cont.)

Note how the candidate has not stated explicitly that she has certain soft skills, but has alluded to them indirectly. The first paragraph highlights the candidate's team working skills, the second her presentation skills, and the third her problem-solving skills and her ability to work to tight deadlines. Clearly, it is not necessary to write about your soft skills for every work or educational experience you have had, just four or five examples highlighting different skills should be sufficient. You can emphasize these skills again in your cover letter (12.24).

9.8 Any differences in a resume?

You can include a skills section either before or after your Education section. Given that a resume is generally a one-page document, you will have very little space. Keep it to 2–3 lines by writing as follows:

Skills

Languages: Spanish (native speaker), English (TOEFL). Computer: Microsoft Word, Excel, Powerpoint. Technical: excellent knowledge of QXC calibrators, good knowledge of UYT.

Before being admitted into many US universities to do business / management courses, you may be requested to carry out a GMAT – a graduate management admission test. This test assesses your analytical, writing, verbal and reading skills in English. Go on Wikipedia to learn more.

Summary: Skills

➢ Have sections for technical skills and language skills, but not for communication skills.

➢ Integrate your communication / soft skills into the other sections of your CV and into your cover letter.

➢ Use the most concise format possible, unless you have space that you wish to fill.

10.1 Who cares about my hobbies and interests? They're my business, aren't they?

These are important because they give an idea of your personality. They are unique and give HR people insights into your character that are difficult to demonstrate in the rest of the CV.

You can also include jobs that you have done or still do that do not fit in easily into your Work Experience section because they are not part of your career path (see 8.7). Such jobs include voluntary work, babysitting, working in shops etc.

10.2 What are considered 'positive' hobbies and interests?

Mention things that:

- show you have a social conscience (e.g. voluntary / charity work)
- highlight your leadership skills (e.g. team captain of a sports team, sports trainer)
- demonstrate your communication skills
- are fun and have positive connotations (e.g. salsa dancing, playing the saxophone), or interesting (e.g. acting), or creative (e.g. pottery, short story writing)
- indicate that others consider you to be a responsible person (e.g. babysitting)
- unusual without being strange (e.g. falconry, acrobatics).

A. Wallwork, *CVs, Resumes, and LinkedIn,*
Guides to Professional English, DOI 10.1007/978-1-4939-0647-5_10,
© Springer Science+Business Media New York 2014

10.3   What should I avoid mentioning?

Avoid mentioning things that most people probably do (e.g. *reading, traveling*). Instead, be specific. Rather than *sports*, write *swimming, hockey* etc. If you put *traveling*, maybe say your favorite destinations.

Do not put activities that are political, religious, or contentious (e.g. *hunting, shooting*).

HR people are interested in your ability to work in a team and in your social skills. So, avoid solitary or nerdy activities (e.g. *computer games, collecting stamps*).

Some activities that most people would consider to be very positive and altruistic, such as being a blood donor, may by a company be considered negatively. In the case of donating blood, for example, this may involve having to take time off work for which the company may have to pay you. Paying you for not being at work doesn't usually make companies happy!

Finally, avoid anything that people tend have strong opinions about. For example, people tend to either love or hate board games or role playing games. So these are best avoided.

10.4   How can I use my interests to provide evidence of my soft skills?

Below is what a PhD student in Cognitive Sciences wrote under 'Hobbies and Interests'.

> I love traveling because I enjoy meeting new people and staying in different socio-cultural environments. My interest in foreign languages and cultures has increased my metalinguistic skills in living and working with people from different cultures thus I have no problems to integrate myself socially. I have been doing various kinds of sports (dance, horse-riding, kung fu, capoeira angola) since I was 5 years old. These activities have played an important role in forming my sense of duty and organization and mostly my interest towards new and different cultures and persons.

She has used her interest in traveling, languages and sport to show her social skills and the fact she is interested in a wide variety of activities and thus has a broad skills base. She has exploited this section to give the reader a clearer idea of who she is as a person, and what makes her stand out from the other candidates.

10.5 Should I write a list or a short paragraph?

The candidate in the previous subsection opted for a paragraph and used around 100 words but clearly believed that such information would help her chances of being hired. Obviously, if your CV is very full, you would not have the space to include such information.

So she could have listed her interests as follows:

traveling, living and working in foreign countries, sports (dance, horse-riding, kung fu, capoeira angola)

The advantages of a list are that it:

• takes up much less space

• is quicker to read.

However a paragraph, if you have space, can give the HR person a much clearer picture of who you are.

10.6   What are the dangers of writing a paragraph?

Here is example from a young graduate (a native English speaker) who was looking for a job in teaching.

> I like to spend my free time outside, playing rugby or walking, working in the garden or reading. I've played for the Altrincham rugby team for several years and worked alongside the youth team. I was involved in the founding of the Rugby Society of Reading University. I regularly listen to live music whenever I have the opportunity and enjoy playing the guitar. I have a keen interest in travel, and spent 7 months after completing my academic studies fulfilling my dream of travelling the world. I visited China, Singapore, Australia, New Zealand and the United States of America during my trip. I feel this trip opened me up to experiences and empathy which would be unattainable in Britain and as such have made me a better rounded person. While away on my travels I spent time working on farms in New Zealand; I found this work gave me good perspective on all the types of work in the world. In my life I have spend a lot of time working in groups or teams, my personality is well suited to the group environment. Groups I have been involved with have helped organise holidays for the disabled, music festivals for charity as well my personal achievements of climbing Mt. Kilimanjaro and working in Tanzanian School.

The problem with writing so much is that you:

- have to use complete grammatical sentences, and thus probably make more mistakes (even this native speaker makes a mistake: *I have spend* rather than *I have spent*)
- may be prone to waffle (i.e. not be very precise or concise)
- use up a lot of space.

You also need to ensure that this section is organized clearly, just as in the other sections.

By writing so much, you are assuming that the hirer has the time to read all that you have written. However, it may depend on who the hirer is. If you are looking for a job for a small organization, and the job requires a candidate with an interesting personality and with good communication, then the hirer may be truly interested in reading about all your personal activities.

10.6 What are the dangers of writing a paragraph? (cont.)

So, again a list might be a better option. Below is an example of how the above candidate could have listed his interests. Note how it is divided into mini sections, which gives the idea of an organized mind and who also wishes to communicate information in the clearest and simplest way possible.

Sports and outdoor activities: rugby (I play for my local team, and founded a rugby society at university), walking, gardening.

Traveling: I have visited China, Singapore, Australia, New Zealand and the United States of America, Tanzania (where I climbed Mt. Kilimanjaro).

Volunteer work: organizing holidays for the disabled, music festivals to raise money for charity (I also play the guitar).

By writing a list you will make fewer mistakes. You can also provide interesting details about yourself that provide evidence of your soft skills, e.g. the parts in brackets in the list above.

For more ideas on writing about your personal interests see 14.18.

10.7 Are there any other tricks for gaining the hirer's attention through my Personal Interests section?

There are two simple tricks you can use:

Firstly, read the job specification carefully and see if there are any non-technical requirements that you could somehow insert into your Personal Interests section.

Below are three extracts from a job specification as a patent examiner.

> The job of a European patent examiner demands a unique combination of scientific expertise, analytical thinking, language skills and an interest in intellectual property law.
>
> You should have a genuine interest in technology, an eye for detail and an analytical mind.
>
> Applicants must also be willing to relocate to Munich, The Hague or Berlin.

Through your interests you could try to provide evidence that you have the skills required. For example, someone who paints or has reviewed papers will have *an eye for detail*, someone who has traveled frequently and who has preferably been to Germany or the Netherlands shows that they wouldn't have a problem to *relocate* and will probably have *language skills*, and someone who designs and makes their own model planes probably has *a genuine interest in technology, an eye for detail and an analytical mind*.

Secondly, if you know exactly who is going to read your CV, then you can find out this person's personal interests on LinkedIn, Facebook, or on their blog or personal website. If you find any genuine matches between your interests and their interests, then you can mention them. But at the interview do not mention that you looked at your interviewer's web pages!

10.8 Any differences in a resume?

Personal interests are not frequently found in a resume. Given the importance of revealing something about your personality, you can put a link to the Interests section of your LinkedIn profile (see 14.8).

Summary: Personal Interests

- Ensure you have a Personal Interests section—this enables interested HR people to get a clearer picture of you as a person. In any case, the uninterested HR person can ignore this section if they wish.

- Only mention activities and interests that have a positive connotation for the majority of people.

- Only use a paragraph if you have excess space. Otherwise use a list, preferably divided into three or four mini subsections.

- Exploit your interests to highlight your soft skills and the skills requested in the job specification.

11 REFERENCES AND REFERENCE LETTERS

11.1 What is a referee?

In the context of a CV, a referee is someone you have worked for or collaborated with, and who can provide an objective appraisal of your technical and social skills.

11.2 Do I need to provide the names of referees on my CV?

In most Anglo countries it is customary to provide the names of referees. This enables potential employers to contact your referees to:

- check that you are who you say you are, and that you have done what you say you have done
- learn more about your personality and skills

HR use references as part of the screening process of candidates.

So, yes, you do need to provide references.

A. Wallwork, *CVs, Resumes, and LinkedIn,*
Guides to Professional English, DOI 10.1007/978-1-4939-0647-5_11,
© Springer Science+Business Media New York 2014

11.3 Where should I put my referees on my CV?

At the end of your CV have a separate section entitled 'References' in which you list three or four people. Provide the following information:

- name
- their relationship to you
- where they work
- their email address (so that the HR person can contact them)
- their website (so that HR can learn more about them)

For example:

> Professor Pinco Pallino (my thesis tutor), University of London, p.pallino@londonuni. ac.uk, www.pincopallino.com

> Professor Zack Madman (in whose lab I did a 3-month internship), University of Harvard, z.madman@harvard.edu, www.harvard.edu/madman

You do not need to write any more than above. For example, the referee's postal address and telephone number are redundant, as it is highly likely that first contact will be via email. The other problem with providing more information is that it can take up a lot of space as highlighted in the examples below. If you provided five references using this format, it would occupy 15 lines – far too much space.

> **Clare Henley (CELTA Tutor):** Manchester Academy of English, St Margaret's Chambers, 5 Newton Street, Manchester, UK, M1 1HL. Tel – +44 (0)161 – 237 5619 Email – chenley@manacad.co.uk

> **Jo Bloggs (Self Help Services Volunteer Coordinator):** Self Help Services, Zion Community Resource, 339 Stretford Road, Hulme, Manchester, UK, M15 4ZY. Tel – +44 (0)844 477 9971 Email – j.bloggs@selfhelpservices.org.uk

11.4 Can I simply say 'references available upon request' at the bottom of my CV?

Yes, but only if you are short of space.

If you have the space available, then there is certainly no harm in putting references. Such references add credibility to your CV. They imply that you have no problem with potential employers doing background checks on you.

Alternatively, you could put the references in your cover letter.

11.5   Will HR people and recruiters contact my referees?

It depends on where in the world they are located. Many HR people and recruiters in the USA and UK and other Anglo countries will contact your referees typically via email but also via phone. A typical email they might write to your referee is:

Re Adrian Wallwork

The above named student has applied to our Department for admission to a Postgraduate Programme of Study (PhD) and has given your name as someone who can inform me of his ability to undertake advanced study and research leading to a higher degree in Physics.

Would you please let me know, in confidence, your opinion of Mr Wallwork's ability, character and capacity for postgraduate study.

Thank you in advance for your cooperation.

11.6   What is a reference letter?

A 'reference' is a letter written by a referee, i.e. the person you worked for or collaborated with – typically your professor / tutor and people you have worked for / with during an internship. Or if you have work experience, then the letter could be written by a past employer.

In this letter the referee gives a brief summary of your technical skills and also your personality (how motivated you are, how easy you are to work with, how proactive you are etc).

11.7 How important is the reference letter?

It is extremely important.

You create your CV and cover letter, so the picture that you present of yourself is naturally very subjective. Also, when you describe your soft skills, you are describing these skills from your own point of view and it is very difficult for you to prove within the CV and cover letter that you really have these skills.

A reference letter is written by a third party who generally has no vested interest in you getting a particular job or not. The role of this third party is to present an objective view of you and your personality.

For the HR person, a reference letter is a kind of guarantee that you are who you say you are.

Note: To learn how (and how not) to write about soft skills, see 9.7, 11.16 and 12.24.

11.8 Who should I ask to write my reference letter? Can I write it myself?

Whenever you work / collaborate with someone in a lab or a company, get a written reference from someone there. You can then use these references as and when you need them. Also, get permission from these people to put their names, position and email addresses on your CV.

Many of the people you ask for a reference might appreciate it if you write the letter yourself and then submit it to them for their signature and approval. This saves the referee a lot of time.

Writing the letter yourself has several advantages. You can decide:

- the exact content
- the structure
- the length
- what to emphasize and what not to mention

But remember, if you write the letter yourself, you must submit it to the relevant person for their approval and signature (see 11.14).

In some cases, your reference person may be contacted directly by the company or institute where you have applied for a position and the employer may send a form to fill out. It makes sense for you to fill out the form with your professor, or at least to request that you be able to see the form before it is sent back to the employer. If you can, get hold of such forms in advance, and plan carefully how to answer the questions.

11.9   How should I ask someone to write a reference letter for me?

Ideally you should ask the person while you are still working for him / her. Then they will have time to prepare the letter before you leave. You can say to this person:

> I was wondering if it would be possible for you to write me a reference letter.

If you decide you would prefer to write the letter yourself (11.8, 11.14), then you can add:

> If you like, I can write the letter myself and then submit it to you for approval.

If you are not dealing with your referee face to face but via email, you can write:

> Dear Professor Smith
>
> First of all I just wanted to say how useful I found the three months in your laboratory. It was particularly useful because ...
>
> I was wondering if it would be possible for you to write me a reference letter.
>
> I am applying for a position at ... / I am applying for a job as a ... at ... It would be much appreciated if you could write the letter by the end of next week.
>
> Alternatively, for your convenience, I could write the letter myself and then submit it to you for your approval and signature.
>
> Best wishes

The structure of the above email is:

1. mention the time spent at the referee's department, institute, company etc

2. add some details of the usefulness of the time spent there

3. ask for a reference letter

4. say why you need the letter, i.e. the position you are applying for

5. give a deadline for the receipt of the letter

6. suggest that you write it yourself

11.10 How important is it for the reference letter to be written in good English?

Very.

A reference letter that is full of mistakes, even if it was not written by you, may undermine what it says about you.

So if you write your own letter (11.8, 11.14), then you can also have the letter checked by a native English speaker to make sure there are no grammatical mistakes. Remember that the level of English of the person who you ask to be your referee may not be very good.

In any case, even if your referee writes your letter, you can still have it checked by a professional. You can then resubmit it to your referee for them to sign again. You may find such a procedure embarrassing but you can write an email such as:

> Dear *name of referee*
>
> Thank you so much for writing me the reference letter, I very much appreciate it.
>
> A native English speaking friend of mine happened to read it and noticed a couple of mistakes.
>
> Attached is a revised version. Would you mind signing it again for me.
>
> Thank you for your help.
>
> Best regards

11.11 Example of a poorly-written letter and a well-written letter

Below are the first two paragraphs of a letter written for a student, Melanie Guyot, by her university professor. As you read the first sentence, try to understand who *teacher* refers to – Melanie or her professor? What other mistakes can you find?

Letter in support to the candidacy of Melanie Guyot as an XXX at the XXX

I am writing this, having assessed the capacity of Melanie Guyot both as a teacher of Medical Robotics (Master of Science in Biomedical Engineering at the University of XX) and during her work as a majoring in biomedical engineering at the Biorobotics Institute, Creative Design area of which I am the coordinator. Also in this last year I got to know the skills and potential of the candidate as a research fellow at our institute.

During the master thesis, I could appreciate its intuitiveness in solving real physical problems that helped to efficiently design an hydraulic actuation system for minimally invasive surgical instruments. She has been able, in that context, to greatly expand her knowledge, showing excellent ability to work in multidisciplinary field of research.

The first sentence contains 52 words in a long series of subclauses. It is very poorly structured and includes ambiguity (who is the *teacher*?). In the second paragraph there are numerous mistakes in the English:

during the master thesis = during her Master's

its intuitiveness = her intuitiveness

an hydraulic = a hydraulic

she has been able = she was able

work in multidisciplinary field = work in a multidisciplinary field

The result is that the referee's poorly written letter reflects badly on the candidate. The reader (HR; recruiter) might associate the referees' poor English and structure with unreliability and lack of professionalism. This lack of professionalism might then, unfortunately, be transferred to the candidate as well.

Below is an example of a well-written letter.

11.11 Example of a poorly-written letter and a well-written letter (cont.)

Adele Tulloch

I am a Professor of Information Engineering at the University of Manchester and I have great pleasure in writing a reference for Adele Tulloch. I first came into contact with Adele when she attended two courses in Computer Networking that I give to Master's students in Computer Engineering - her scores in the examinations were excellent, i.e. way above the average.

After completing her Master's, Adele won a 5-month research scholarship targeted at the bandwidth estimation of the bottleneck link along an Internet path.

During this time Adele worked in my research group. Adele showed no difficulty in absorbing new knowledge and in gaining the needed knowhow. She was both proactive and creative, often producing truly original and outstanding results (see attached list of papers).

Adele not only has excellent technical skills, but has several other qualities that I find to be quite rare in someone of her age. Her high level of independence meant she was able to carry out the work by herself with only occasional guidance from me. I found her very receptive and easy to talk to. She had a useful knack of being able to take my ideas to the next possible level and suggest possible directions of further investigations. She always managed to meet any deadlines I set her and always with a self-critical eye on her own work.

In summary, I have absolutely no hesitation in recommending Adele Tulloch as a xxxx. Please feel free to contact me should you need any further details.

Best regards

Prof. Adrian Wallwork

The above letter will create a positive impression on the reader because it:

- is well organized (see 11.12)
- gives a clear indication of both Adele's intellectual and technical capacity as well as her personality
- begins and ends in a very positive way, thus creating a good impression both at the beginning and the end of the letter

You can find another good example in the next subsection.

11.12 How should I structure my reference letter?

Your reference letter needs a very clear structure which will highlight your key skills (both technical and soft) and will thus act as an objective support to what you have written in your CV. Here is a possible structure:

0) Heading

1) Positive opening sentence

2) Referee's position

3) Referee's connection to candidate

4) Details about candidate's qualifications

5) Reference to candidate's wonderful personality

6) Positive conclusion

7) Salutation

Now let's look at how the poorly written letter in 11.11 could be rewritten. Each paragraph of the letter is an example of the structure above.

Melanie Guyot [0]

It is a pleasure for me to have the opportunity to thoroughly recommend Melanie Guyot [for the position of ...] [1]

I am the coordinator of the Biorobotics Institute at the University of Monpellier. [2]

I was Melanie's supervisor while she was doing her Master's of Science in ... She was also a student in my class on medical robotics. [3]

During her Master's thesis, Melanie demonstrated great intuitiveness in solving ... In fact, she played a major role in ... She also ... [4]

Melanie has a bright and lively personality and works extremely well in teams, both as a team member and team leader. She showed a clear demonstration of these skills when ... [5]

I very much hope that her application will be taken into serious consideration as I am sure that Melanie Guyot represents an excellent candidate. [6]

Best regards [7]

Pierre Lepoof

(pierre.lepoof@institute.com)

11.12   How should I structure my reference letter? (cont.)

The letter above highlights the following points.

- There is no initial salutation, i.e. the letter is not addressed to anyone in particular. In fact, the letter is intended for anyone that Melanie chooses to write to.

- The heading of the letter simply contains the candidate's name, rather than adding the position that candidate is looking for on one specific occasion. This means that you can use the same reference letter for more than one job application.

- Each paragraph has a specific purpose. Also, by dividing the letter up into multiple paragraphs (rather than one or two long paragraphs), the reader will be able to quickly locate the information he / she is interested in.

- In Paragraph 2, the referee first establishes his / her credibility. The HR person needs to know that the reference is written by someone who is qualified to write such a letter. Then in Paragraph 3, the referee answers an HR person's question: How does the referee know the candidate?

- Paragraph 4 will probably be the longest paragraph in the letter, and may even be divided up into more than one paragraph. This is the part of the letter where the referee tries to establish the credibility of the candidate, and gives objective support to what the candidate has written in his / her CV. Further support is provided in Paragraph 5, which outlines the candidate's social / soft skills (see 9.7 and 12.24).

- When writing any kind of letter or email, it is good practice to begin and end on a positive note (Paragraphs 1 and 6).

- The final salutation is brief (*Best regards*). There is no need to write 'Feel free to contact me should you need further details'. The whole point of a reference letter is that the referee implicitly gives their permission to be contacted, so there is no need to state this. The aim is to keep the letter as concise as possible.

- The referee provides their email address (if this is not already on the header / footer of the letter) so that they can be contacted.

11.13   How should the reference letter be laid out?

The reference letter should be printed on headed paper (i.e. the paper of the company or institute where your referee works). This means that the referee's address will be contained either in the header or the footer.

So, a clear simple layout for a one-page cover letter is:

Date [example: 10 March 2020]

one or two lines of white space

Heading [in bold and centered]

Paragraph 1

Paragraph 2

Paragraph 3

Paragraph 4

Paragraph 5

Paragraph 6

one line of white space

Salutation

Name of referee

Referee's contact details [possibly in brackets]

Note that to make the letter as easy as possible to read:

- everything is aligned to the left (apart from the heading)
- there is no indentation
- there is a 6 pt space after each paragraph

11.14 Is it acceptable and ethical for me to write my own reference letters? What are the dangers?

It is both acceptable and ethical to write your own reference letter, providing that your referee:

- sees the letter and approves the content
- checks that his / her contact details are correctly written on the letter
- signs the letter

But be aware of the dangers. Look at the beginnings of these letters. What do you notice about them?

Shanghai, December 3rd, 2020

To whom it may concern,

It is a pleasure to write this letter of reference for Dr. Fu Hao.

From December 2019 to July 2020 Ms. Fu Hao did her research work for her bachelor thesis in my group, supervised by Associate Prof. Zhaobin Yang. Due to her interests on "molecular design", she was working on ...

Shanghai, January 16th, 2021

To whom it may concern,

It is a pleasure for me to write the reference letter of Dr. Fu Hao for her application of postdoc position.

In March 2017 Ms. Fu Hao started her PhD study in the field of Chemistry and Physics of Polymers at Shanghai Jiao Tong University (SJTU), China, under my supervision.

Bogota, December 3rd, 2020

To whom it may concern

I am pleased to write this letter of recommendation for Dr. Fu Hao.

I met Fu Hao in 2018, during a Conference in Shanghai where she attended a session in which I was a speaker. After that she contacted me asking to join temporarily my research group during her last years of graduate studies pursuing a PhD degree in Chemistry at Shanghai Jiao Tong University.

She joined my research group at the end of 2018 ...

11.14 Is it acceptable and ethical for me to write my own reference letters? What are the dangers? (cont.)

The problem is that they all look the same both in terms of layout and content:

- the date is expressed using the same format

- the location of where the letter was written is always indicated (though this is not a recognized standard in Anglo countries)

- they all begin with *To whom it may concern*

- the opening sentence is very similar

- the name of the candidate is referred to in the same way – the first time as Dr and the second time as Ms

This may make the HR person suspect that the candidate has written her own reference letters, and that the referees themselves may actually have not seen the letters. So if you do write your own letter then try to use a slightly different format. You can modify the:

- date e.g. 10 March 2020; March 10th, 2020; March 10, 2020

- layout

- font

- use and non-use of paragraph indentation

- order that information is given

- salutations

- overall length of the letter

11.15   Is it OK for the reference letter to include negative information?

Yes. You need to be honest. And it also makes you sound more credible.

In the example below, the possibly negative information is highlighted in italics. Note that although this info might initially be interpreted as negative, a logical explanation is given.

> Once in Bogota she readily accepted to take up a new research topic (fluorinated latex blends for nanostructured coatings) that had been in stand by because of lack of specific financing, *only vaguely related to her previous research* but well within her already broad expertise and skills. In the following three years she actually did all the research work and achieved the results that allowed her to obtain her PhD from Jiao Tong University at the beginning of 2020.
>
> In the meanwhile, I had her work also on other projects that *did not allow her to write papers* due to confidentiality, but allowed me to finance her stay in Bogota after an initial period in which she had been supported by her University. These are the main reasons for the apparently long time it took her to complete her PhD and for the limited number of papers published.
>
> *Unfortunately I could not extend her contract* because of lack of adequate financing, but I am confident that with her good background, skills and expertise she will very quickly find a new postdoctoral position within an excellent research group.

The example above highlights how a cover letter can be used to explain any apparent anomalies in the CV, such as why you:

- apparently changed the course of your career (i.e. a seemingly illogical sequence of jobs or research positions)
- failed to publish the expected number of papers
- had to interrupt a particular internship

The same letter also contained these two comments:

> Her approach to scientific problems and to the interpretation of the experimental results is *still a little too focused on details,* however she is making great efforts to look at the broader picture.
>
> ....
>
> She is effective in reporting her results, *though she still requires some support* in the initial and final stages of writing a paper. For the latter issue, I believe I was not the best possible supervisor because I tend to always cover the final stage myself.

Having such negative information (which in this case is not too serious) makes you seem like a normal human being. Having a letter that is 100 % praise may sound less credible – does this candidate really have no faults? do I want such a perfect person on my team?

11.16 More examples of typical things mentioned in a reference letter

Writing about your own or someone else's qualities and soft skills (see 9.7) is not easy. Below are a few examples to give you some idea of the kinds of things that can be written about these skills. The examples also highlight the many areas that a reference letter can potentially cover.

HOW REFEREE KNOWS THE CANDIDATE

I have known Valentina Putin since November 2028 when she joined our organisation, and have always found her a most pleasant and trustworthy person to work with. She has a friendly and outgoing personality and is capable of flashes of brilliance in her work.

DESCRIPTION OF CANDIDATE'S RESPONSIBILITIES

From late 2029 Ms Bloggs was involved in coordinating project development and was appointed Projects Manager in Feb 2031 to oversee work projects involving over 150 people. The Project Management duties were as diverse as the projects themselves, which varied from Community Theatre, Visual Arts Displays and Electronic Workshops. Managerial control and operations involved personnel activities, budgetary and financial administration, research and development ...

RELATIONSHIP BETWEEN REFEREE AND CANDIDATE

As a foreign PhD student with unusually good spoken English and communication skills, he quickly established an excellent relationship with the members of my small research group and of my department. He was effective when I asked him to tutor undergraduate students, and showed from the very beginning a great deal of independence, good understanding of the subject as well as initiative and critical thinking (in fact, I even had heated discussions with him in a few occasions, when I was expressing different ideas on how to proceed with the experiments or what to write in the papers; I should mention that in the end he often turned out to be right!).

HOW WELL CANDIDATE WORKS IN A TEAM

Her excellent teamworking attitude allowed her to collaborate effectively with other researchers and to quickly pick up the essentials of new characterization techniques. With clear targets and good motivation she can be a highly dedicated worker.

CANDIDATE'S PERSONALITY

As a Projects Manager, Herman showed himself to be a reliable and hardworking person with a capacity for original ideas as well as close attention to details. He is good at organising work and got on well with people around him. He proved capable of working to tight schedules and maintained a good control over the Project Finances.

CANDIDATE'S HEALTH RECORD

During the period of her employment, I have known her to be very resilient and healthy individual. There is nothing in her health record to prove otherwise.

11.16 More examples of typical things mentioned in a reference letter (cont.)

CANDIDATE'S OUTSTANDING SKILLS

Gojko's role in both projects during the design and implementation phases was absolutely fundamental in making them successful. His extraordinary ability in working out both management and design problems as the projects developed was immediately clear to all the participants.

HOW SUITED THE CANDIDATE IS FOR THE JOB APPLIED FOR IN INDUSTRY

From the duties and responsibilities detailed in your job description I can closely identify Ms Yamashta's abilities and potential to be relevant the requirements. She has an analytical mind and strength of character to meet the demands of such as a post.

HOW SUITED THE CANDIDATE IS FOR THE JOB APPLIED FOR IN RESEARCH

Dr. Veena Huria is ideally suited for this grant. She was a visiting scientist in my laboratory for the period of June 15 to August 15, 2099. During this first visit, the work accomplished by Dr. Huria was very impressive, and her attitude toward her work was both proactive and refreshing. Four publications in international journals have resulted from work: two published and two in preparation.

FOUR EXAMPLES OF CONCLUDING SENTENCES

I can wholeheartedly recommend Ms Kanjika. She is a very willing person and I have no doubt that she will be an asset to your organisation.

Vladimir has my full support in his application to you and I am of the opinion that his considerable potential will be an asset in your company.

In conclusion, I can highly recommend Hao Pei Lin not only on a professional level, but also on a personal level. She is, in my opinion, very professional as well as being an outstanding sales person.

Such is my appreciation of Fernando's work and skills that I would very warmly recommend him for the position he is seeking in your company.

11.17   Any differences in a resume?

References are not typically found on a resume. But if you have space,
I see no harm in putting them. If you decide not to put them, you can refer
to them in your cover letter.

Summary: References and reference letters

- Have a separate section for references at the bottom of your CV, or if you have no space mention them in your cover letter.

- A reference letter increases your credibility as a suitable candidate as it is written by a (presumably) objective third party who has had direct experience of working with you and who can substantiate both your technical and soft skills.

- Collect reference letters for every job or project that you work on.

- Get permission from your referee to put their contact details on your CV.

- Consider writing the letters yourself and then submit them to the referee for approval / modification.

- The letter must be perfectly written both in terms of content, organization and English.

- Typical structure: 1) heading 2) positive opening sentence 3) referee's job 4) referee's relation to you 5) details of your qualifications 6) description of your personality 7) positive conclusion.

- The letter can include negative information about you, but the emphasis should always be on the positive.

- Use a simple layout with everything aligned to the left.

- Print on good quality paper.

12 COVER LETTERS

12.1 What is a cover letter? And how important is it?

A cover letter is the letter that you send along with your CV to a potential employer. A cover letter can be a printed document or an email.

The main aim of a cover letter is to convince the reader to look at your CV. You can do this by giving the HR person the impression that by hiring you their company will improve its efficiency, production and sales, or that you will contribute unique knowledge to a research group.

There is a chance that the recruiter will not even read your cover letter or they may read it after the CV.

However, if they do read the letter first and they don't like what they read, they probably won't look at your CV. Thus, the cover letter is extremely important.

When you are applying for a job online, you may not need to write a cover letter.

12.2 What is a motivational letter? What is a statement of interest?

There is no real difference in the aim of a cover letter, motivational letter or a statement of interest. All of them are opportunities for you to expand upon some of the more salient and interesting points of your CV and to interpret their significance for the HR person.

These letters are a way for you to sound dynamic and really differentiate yourself from other candidates. Your letter should answer such questions as: What targets did you reach? How well did the projects go? What did you learn from them? How could this experience be applied to the position you are applying for?

From an employer's point of view, these letters / statements are demonstrations of

- how much you care about getting the job

- your writing ability

A. Wallwork, *CVs, Resumes, and LinkedIn,*
Guides to Professional English, DOI 10.1007/978-1-4939-0647-5_12,
© Springer Science+Business Media New York 2014

- your attention to detail

- your communication skills

The key differences are that a motivational letter or a statement of interest tends:

- not to be in response to a specific job advertisement

- to be used in academia rather than business

- to be longer than a cover letter

There is no real difference between *motivational letter* and *statement of interest* – they mean the same thing.

So, if a research institute or university requests a motivational letter or a statement of interest, you can follow all the suggestions in the rest of this chapter, but simply provide more details. This means that you are likely to write more than one page of text.

A website with very good suggestions for a statement of interest / motivational letter is:

> http://www.fordschool.umich.edu/downloads/writing-effective-statement-of-interest.pdf

For an example of a long cover letter that also functions as a motivational letter see 12.37.

12.3 What does a recruiter expect to find in a cover letter?

The recruiter wants to know:

- what job you are applying for (and perhaps where you saw the advertisement); alternatively, who gave you their name

- why you are interested in this field / company

- how your skills and experience directly apply to the advertised job

- the benefits for the company / institute of employing you

- that you know something about the institute / company

- how you might fit in both in terms of your skills and your personality

The recruiter wants to receive this information as fast and as easily as possible. This means that your letter should be well laid out and organized, and should be short and concise.

12.4 Can I use the same cover letter to several companies / institutes?

No.

Your CV and your cover letter should both look as if they have been written for a specific company or institute. This only entails changing a few details so that your qualifications are a better fit with the requirements of the employer.

In any case, be very careful when using the same cover letter for several companies – ensure you change the address, date, name of person you are writing to, and any references to the company / institute.

Never send a photocopy of your cover letter – each letter you send should be printed separately.

12.5 How can I make sure that someone actually reads my cover letter?

If you are applying for an advertised job, then recruiters will be motivated to open your email or envelope and read the contents because it is in their interest to find a good candidate.

However, if you are not applying for a specific post, but you are writing to an institute or company in the hope that they might be interested in your qualifications and skills, then you have to follow a strategy (see 12.10– 12.19).

12.6 What subject line for an email should I use for an advertised position?

Your subject should include:

1. the position you are applying for

2. where you saw the advertisement (put this information in brackets)

Examples:

Application for post-doc researcher (LinkedIn)

Internship in Prof Smith's Lab (ad on your website, 7 Mar 13)

The reason for you putting where you saw the job advertised is purely for the company's own internal statistics. It helps them to decide the best location for their adverts.

12.7   What should I write on the envelope?

If you are writing by traditional mail, then the envelope could look like this:

> Helen Smith
>
> Human Resources Manager
>
> ABC Inc
>
> 22-66 The Boulevard
>
> Boston
>
> MA 02122

Application for Sales Assistant

You would probably say in the cover letter itself where you saw the advertisement.

12.8   If I am sending the cover letter by traditional post, should I type the letter and envelope or write them by hand?

Most companies would probably expect you to type the letter and envelope.

However, some companies may specifically ask you to write by hand, as they may want to use a graphologist to interpret your personality from your handwriting.

If you receive no specific instructions from the company, then type your letter and envelope. But if you have particularly nice (but easily readable) handwriting, then you might like to write by hand.

12.9   What are the problems in applying for job that has not been advertised?

Encouraging the recipient to open and read an unsolicited email is one of the biggest problems of working people around the world.

You cannot assume that just because you send someone an email that they will actually open it. No one is under any obligation to respond to an email (or letter), so there is no point in getting frustrated if you receive no reply.

Given that many people in companies receive hundreds of emails a day, there is a chance that your email will simply not get noticed or it may even go directly into the recipient's spam. Even if it is noticed, it may not be opened as it will be at the bottom of the recipient's list of priorities.

This means that email subject lines such as the following are not likely to encourage the recipient to open them:

Info on job positions

Do you have any sales positions?

Application for the position of junior developer

Also, even if they are opened, the recipient may not be motivated to act on the email simply because it requires extra work for them.

Subsections 12.10–12.19 explain how to increase the chances of your email being opened and read.

12.10 What subject line should I use when applying for a job that has not been advertised but where I know someone who already works for the company / institute?

Think about why you open certain emails but not others. Generally you open an email when it is clearly for you, for example it comes from someone you know or it is obviously about a work issue that regards you.

You are probably reluctant to open an email from someone you don't know. This could be because you are worried it might contain a virus or simply because you have better things to do.

Faced with this problem, you need to find a way to get your recipient to open your email, even though the recipient has no idea who you are.

Imagine you want a job at ABC. You know someone else who works at ABC, her name is Xun Guan. Your strategy could be:

1) Email Xun Guan and ask her who is the right person to contact at ABC. Xun Guan tells you that Kay Jones is the right person.

2) Email Kay Jones with this subject line:

Xun Guan: Assistant Marketing Manager position

The idea is that Kay will see Xun Guan's name in the subject line. This will give Kay confidence to open the mail, and she will see that your first line is:

> Dear Kay Jones, Your name was given to me by Xun Guan, who thought you might have a position available for me as an assistant marketing manager.

In any case, Kay may be able to see the beginning of your email without actually opening the email itself (i.e. in the preview pane). This is why the first words of your email are so important, because depending on the effect these words have on the recipient, he / she will decide whether to open your email or not.

Of course, there is a chance that Kay Jones does not even know Xun Guan – the company may be very big. In this case your subject line would be:

> Xun Guan (ABC, marketing dept): Assistant Marketing Manager position

And your first line would be:

> Dear Kay Jones, Your name was given to me by Xun Guan, who works in your marketing department. He thought you might have a position available for me as an assistant marketing manager.

If you were applying for a job in academia, then the strategy is the same (though you would probably exploit your professor's personal contacts first). In this case, contact someone who already works in the research team where you would like a position. Given that the research team is likely to be relatively small compared to a company, you will not need to explain to your recipient who your contact person is. So your subject line and first line of the body of the text could be:

> Xun Guan: Post-doc position in your team

> Dear Professor Gomez, Your name was given to me by Xun Guan. I was wondering whether you might have a position available for ...

12.11 Is it not a strange solution to put a third party's name in a subject line to someone who I don't even know?

Yes. However your only aim is to get the recipient to open your email. There is nothing devious or unethical in this solution, and the recipient will not be angry with you for adopting this approach.

Also, this approach works. I have used it myself and I have recommended it to my students, with positive results.

12.12 How can I use LinkedIn members when applying for job that has not been advertised and where I do not know anyone in the company / institute?

Finding a job that has not been advertised and is in a company or institute where you do not know anyone is a much more difficult task than those outlined in the previous subsection.

The best solution is probably to telephone the company and institute directly (see 12.16).

If you don't feel confident enough to use the telephone, then you can adopt the same strategy as in 12.10–12.11, but with an additional preliminary step.

First you need to find the name of someone who already works in the company or institute where you would like a job. The best way to do this is through LinkedIn. Go through your contacts to see if anyone in your first-level contacts works in your chosen company. If not, then you need to identify a second-level contact.

The way to find someone is to use LinkedIn's search engine and type in the url of the company or institute.

If you want a job in a company, you just need their url e.g. ibm.com. You can also narrow this down by country, e.g. *ibm.de* (where *de* stands for Deutschland, i.e. Germany). At the time of writing, the search engine is not perfect. For example, even if you put your search items in inverted commas (e.g. "ibm.de") the engine will return LinkedIn members whose profile contains not necessarily *ibm.de* but *ibm* and *de* as separate words (*de* is a word in some languages).

Regarding jobs in research, let's imagine you want a position at the University of Bologna in Italy. First you need to find out the url for this university, which is unibo.it (where *uni* stands for university, *bo* for Bologna, and *it* for Italy). Presumably you want a job in a specific department, so you would also need to find the url for that department, for example the engineering department: ing.unibi.it (where *ing* stands for *ingegneria*, i.e. *engineering*).

If you then manage to locate someone in your network and whose position in the hierarchy is similar to or lower than yours, then you can send an InMail to them. To a first-level contact you can write:

> Hi, I am looking for a job in your company / research team and I was wondering if you could help me. Do you know who would be the best person to contact? Specifically, I am interested in a job in ... Thanks very much for any help you can give me.

To a second-level contact, the text is the same but with a small variation in the first sentence:

> Hi, Your name was given to me by a mutual LinkedIn member – Kamran Dehkordi. I am looking for a job …

You can then proceed as in 12.10. For more on the benefits of LinkedIn see Chapter 14.

12.13 How should my cover letter look?

Find documents written by the institute / company and imitate their:

- style
- layout – e.g. use of white space
- font and font size

Make it seem that you already work for that institute / company!

12.14 How should I address the recipient?

How effective are these?

> To whom it may concern
>
> Dear Sir / Madam
>
> For the attention of the human resources manager.
>
> To the head of the Risk Analysis Department

When you receive an email or letter, do you prefer to be addressed anonymously (as in the examples above) or to be addressed by your name?

Generally speaking, you will get a much better reaction if you address someone by their name. So, find out the name of HR person and address him / her directly e.g. *Dear Hugo Smith.*

The fact that you have taken the trouble to find out the name of the HR person will show that:

- you want the job more than the other candidates (you are differentiating yourself from the other candidates who have not made the same effort as you)
- you are proactive

Both of the above are qualities (i.e. determination, proactive nature) which in any case the company or institute will be looking for in a candidate.

12.15   How can I find out the name of the right person to contact?

You can find out through a third person (e.g. via LinkedIn as outlined in 12.12) or you can telephone the company / institute directly. The problem of using the company or institute's website is that it may not have the most up-to-date information, and so you risk sending your letter to someone who no longer works there.

Finding out the right person to contact is more important when you are applying for a job that has not been advertised. If, on the other hand, it has been advertised, then the recipient will be expecting to receive letters and emails and will not be affected as much by seeing their name.

12.16   What should I say on the telephone in order to find out the name of the right person to contact?

Don't be afraid of using the telephone. Even if your English is not very high level, you don't need to say very much and you can practise the phone call before you make it. Here is a typical dialog between a receptionist and a candidate (in italics).

ABC, good morning, how can I help you?

I would like to apply for a job as a technician. Please could you tell me the right person to contact?

The human resources manager is called Tao Pei Lin.

Could you spell that for me please?

Yes, that's T-A-O and ...

T-A-O. Sorry, could you speak a bit more slowly please?

Sorry, yes, so it's T-A-O and then P-E-I, and then new word L-I-N.

P-E-I?

Yes. Then L-I-N.

L-I-N. Sorry, can I just repeat back the name?

Of course.

So it's T-A-O and then P-E-I and then the last name L-I-N.

Actually Tao is her surname and Pei Lin is her first name.

OK, thank you for your help. Goodbye.

Goodbye.

Clearly, it is essential to ensure you have got the correct spelling of the name (not all names will be as potentially difficult as the one given in the dialog!). Now that you have the name, you can look up the person on the company website to get her email address. Alternatively, if you wanted to find out the address on the telephone you could say:

Could you just give me his email address?

So, is his address john dot smith at abc to com?

If you are not confident enough to use the phone, then you could always ask a friend to make the call for you.

If the receptionist wants to transfer your call directly to the HR person you can say:

No, it's OK I just wanted her email address.

Alternatively, you might want to take this perfect opportunity to speak to the right person – see 12.17 to learn how.

12.17   What should I say on the telephone to the HR person?

Here is a possible dialog between an HR person and a candidate (in italics).

HR. Tao Pei Lin.

Good morning. I am calling to enquire if you have any positions open for a junior developer.

I am not sure if we do. In any case if you would like to email me your CV, I will pass it on to the right person.

OK that sounds great. My name is Javier Morales. If you give me your email address I will send it immediately.

OK, so it's ...

The conversation is unlikely to be more detailed than the one outlined above, so you do not need to be able to speak amazing English to be able to conduct such a call. The important things to remember are:

- the less you say, the fewer mistakes in English you will make. So keep the call brief. Prepare for the call in advance (decide exactly what you want to say and how to say it) and do not improvise

- make sure you leave your name. This means that your recipient will recognize your name in their email inbox and will thus open your email

- double check the address of the person you are going to email

12.18 What can I do if I have failed to find out the name of the HR person? To whom should I address my letter?

If you cannot find out the name of the HR person (or other relevant person) then you can write:

Attn: Human Resources Manager

Attn: Sales Manager

You write this aligned to the left and one line above your subject line.

12.19 Who should I address my email to?

If you are applying online directly on the company's website, then there is no need to address the email to anyone. So simply begin your email with no salutation. Do not write: *hi, hello, good morning*, etc.

The strategy is to write the least amount possible in order to avoid mistakes. For example, you might think that a initial salutation would be:

Dear Sirs

Ladies and Gentlemen

but the above two examples would be totally inappropriate in an email and would have a negative impact on your credibility. So, if you don't put a salutation, you cannot make any mistakes in English, spelling or level of formality.

12.20 How important is the reader's first impression?

What impression do you think an HR person would have of these beginnings of two cover letters?

1) I know about Your job position and I would like to give my effort to your company, welcoming the opportunity to utilise the knwoledge and experience I have gained ...

2) My name is Nerveena Popovic and I would like to become part in a dynamic and innovative field. I am looking for a stimulating and strongly international atmosphere that favors my career development.

The problems are:

- both candidates have written the letter from their own point of view, i.e. what they want rather than what they can offer the company

- the first candidate has made two errors in his English. He has written *your* with a capital letter. It is not a convention in English to capitalize letters to show respect to the reader. He has also spelt *knowledge* incorrectly. This indicates that he has not taken the time to check his letter, which indicates i) that he is unreliable, ii) that he has totally underestimated the importance of the letter

- the second candidate has begun with her name. Your name will be in your signature, so avoid repeating the same information – keep your letter as short and concise as possible.

These would probably be enough to make the recipient stop reading the letter, with the result that the candidate's CV will probably not even be looked at.

First impressions are massively important. If the initial impact is not 100 % positive, you have probably lost your opportunity.

12.21   What is the typical structure of a cover letter?

Your letter can be organized as follows:

- say what position you want (and where you saw the job advertised)

- say what you're doing now, and when your current position will end

- provide a very brief selected past history that will interest the reader and give you credibility

- show that you know about the company or research team and highlight the benefit for them of having you in their team

- brief ending – further details can be given in the next email. Many CV / resume experts, particularly in North America, recommend using an assertive ending in which you state that you will be telephoning to arrange an interview. They claim that this will show initiative. I disagree. I see no benefit of doing this – it is the company's prerogative to suggest an interview, not yours (see 12.28)

12.22 How should I write a letter for a position in business?

Below is an example of how to apply for a position in business or industry. It also shows a possible layout.

Their address

Date

Application for *name of position of job (this line all in bold)*

Dear *name of person (find out HR from website or ring the company – show that you have initiative)*

I saw your advert for a *name of position* on your website / in The Times newspaper / in ABC Journal.

[*Alternatively, if you have been given the name of the HR person by a mutual third party then you can write*: Patrizia Ravenna, who works in your sales department, has told me that you have a position available as a …]

I think I would be qualified for this position because / I think I may have the qualifications you are looking for because:

- a
- b
- c

I would particularly like to work for IBM because … / The skills I think I could bring to IBM are: *Write any additional things that you have not written in a, b, c above. The idea is to show that you know about the company (IBM) and that you would fit in perfectly with them*

I am attaching my CV along with references from various professors and previous employers. I would be available for interview from June 20 (when I finish my current researcher contract at the University of …)

I look forward to hearing from you.

legible signature

John Smith

A note on punctuation.

After the name of the person you are writing to you can punctuate in one of the following ways:

Dear Adrian Wallwork [*no punctuation*]

Dear Adrian Wallwork: [*a colon, typical of North America*]

Dear Adrian Wallwork, [*a comma*]

Whichever of the above you use, the first word of the following sentence will begin with a capital letter. For example,

Dear Adrian Wallwork,

Your name was given to me by George Clooney, who thought you might be …

12.23   How should I write a letter for a position in academia?

Below is an example:

I am writing to inquire about the possibility of a postdoctoral position in your laboratory.

I graduated in 2020 with an MSc in X. I am currently a PhD student in Prof Y's laboratory at the University of Z and I plan to graduate in June 2013. My PhD work has focused on xyz. All my work to date has been published in articles in top international journals. I have also presented my research at several international conferences and have taught two undergraduate classes.

I have experience in:

- x

- y

- z

I know that you are currently working on yyy, and I believe that my experience in this area (three EU-funded projects) would be an asset to your team.

I look forward to hearing from you.

What the above letter highlights is that the candidate:

- is writing from the reader's perspective rather than her own perspective. This is clear from the fact that she doesn't stress what she wants and how it would benefit her, but rather how her experience might benefit the research team

- shows that she has various soft skills that are essential for the position, i.e. she can write reports, she can teach (and thus should have good communication skills), and she can give presentations
- believes she has the right technical skills for the job
- is familiar with the team and their work
- can express herself clearly and concisely

12.24 How should I mention my personality and soft skills?

HR want to know not just about your technical skills and experience but also need concrete evidence of these skills. They also want evidence that you have good communication skills e.g. an ability to work in teams, and to do presentations.

Would you write the following?

> I hate human relationships. I'm lazy, with a zero capacity for teamwork. I have rarely worked during my studies and I never meet deadlines. It is important for me never to finish my projects.

You wouldn't; so don't write this either:

> I love human relationships. I'm proactive with a high capacity for teamwork. I work hard. I like to complete my projects and goals.

Firstly, the qualities mentioned are obvious. Anyone hoping to get a decent job should have such qualities. Secondly, the candidate has provided no evidence that he / she actually has such qualities.

Instead you need to write something that is specific to you and that really demonstrates these qualities.

Companies are interested in various soft skills, for which you need to provide evidence.

Below are various examples that you could write in your cover letter.

ABILITY TO WORK IN A TEAM

> I have worked in several research teams and also during an internship at IBM in Houston in 2020.

PROFESSIONALISM, PRO-ACTIVE, FLEXIBILITY, MEETING DEADLINES

> During my internship I worked on several very diverse projects and managed to meet all the deadlines.

PROBLEM SOLVING SKILLS

The two projects I worked on involved solving complex engineering problems and I very much enjoyed taking part in brain-storming sessions with the team.

PRESENTATION SKILLS

I presented two of the three projects I worked on to clients, both of whom then went ahead and purchased the product.

ABILITY TO WRITE TECHNICAL DOCUMENTS

My duties include liaising with clients and writing specifications.

Obviously you don't need to mention all of the points above. You need to judge which ones would be the most relevant for the job and which ones you have practical experience of that you can demonstrate in your letter.

When mentioning soft skills, it is essential that you substantiate these skills. In the example below the candidate actually proves that he does not have the skill that he claims to have (i.e. good English).

Even if I'm *not expert* in all programming languages *your* request, I am highly motivated to learn new skills and have a good attitude to relate with customers and to find solutions to complex problems.

Additionally I *have good knowledge of english language,* I'm *spanish mother tongue* and I live in Quito.

The example above is full of English mistakes: *even if* (even though), *not expert* (not an expert), *your request* (you request), *good knowledge of english language* (a good knowledge of English), *spanish mother tongue* (my native language is Spanish). Clearly, not only is the candidate's English poor (rather than good), but he also hasn't even bothered to check his letter. This creates a very bad impression, and his CV is likely to be quickly discarded.

12.25 Using my cover letter, how can I make it look as if I am perfect for the job advertised?

Let's imagine that the advertisement you are responding to contains the following selection criteria.

❑ University degree pertaining to design, implementation and evaluation of environment-related programs and projects

❑ Knowledge of bank operations and institutional issues

❑ Ability to produce high-quality output in response to tight deadlines

❑ Fluent English

A good strategy is to list your skills in the same order (where possible and logical) as in their list of selection criteria. Your letter could be:

Dear Benedykta Kajilich,

I am very interested in the position of ... advertised in ... on 10 March.

I have a degree in financial ecology and have worked as an intern at the Cooperative Bank in England for three successive summers, which is where I also learned to speak fluent English. During university I have been involved in several projects and managed to complete all of them ahead of schedule.

Not only have you shown that you have the right qualifications, but you have also demonstrated your communication (ability to organize information into a format that matches the reader's expectations) and sales skills (you can sell your own experience).

12.26 Is it OK to use bullet points in my letter?

Yes.

12.27 What information do I **not** need to include in my cover letter?

The aim of a cover letter is to make key information stand out and to use the least number of words possible. This means that you do not want to include any information that is not strictly necessary.

You do not need to include the following information which is illustrated by the phrases in italics.

1. Your name at beginning e.g. *My name is Nguyen Hung and I ...*

2. Enclosed CV e.g. *Enclosed please find my CV.*

3. Availability for interview e.g. *I am available for interview at any time.*

4. Contact you for additional info e.g. *Please feel free to contact me should you need any further information.*

5. References upon request e.g. *I am happy to provide references on request.*

In a cover letter your name in the main text is irrelevant because it will be in your signature to the letter and also at the top of your CV.

If you are applying for a job you will obviously attach / enclose your CV, so you do not need to mention this fact.

If you really want a job you will always be available for interview, you only need to make reference to the interview if there are times when you absolutely cannot come (in which case you can write: *I will not be available from June 22 to June 29*; you don't need to explain why). If your potential employer is interested in further information about you or wishes to contact your references, they will ask you – you don't need to tell them that you are willing to provide such information, why would you <u>not</u> be willing?

In addition, don't have more than one salutation (*Best regards* is sufficient).

By not mentioning the above five points, you will reduce the length of your cover letter by up to 25 %. This will also help to make your key points stand out, as you can see from the example below. By deleting certain phrases the letter below is reduced from 168 words to 116, and from 17 lines to 13. Such deletions could reduce a two-page letter to a more standard one-page letter.

Dear ~~Mrs~~ Helen Murray

~~Application for position of~~ Post doc DVB researcher

With reference to the vacancy advertised on your website, I would like to apply for the position of post doc researcher.

~~My name is Tek Saptoka.~~ I am particularly interested in taking active part in the research activities of the Connectivity Systems and Network department. In fact, I spent an internship at XYZ where I also managed a small team of PhD students.

~~As you will note from my CV / resume,~~ In November 2023, I graduated in Telecommunication Engineering at the University of Zulia, Venezuela. For my Masters project, I was an intern at XYZ, where I joined the DVB-TM ad hoc working group and worked on carrier synchronization algorithms for DVB-S2 applications.

~~I am available for interview at any time.~~

~~If you need any further information, do not hesitate to contact me either by phone or e-mail.~~

~~Enclosed please find my curriculum vitae.~~

I look forward to hearing from you.

~~Yours sincerely,~~

Tek Saptoka

12.28   What should I avoid writing in my cover letter?

The previous subsection outlined information and typical phrases that are redundant in an email. However, none of them by itself is likely to prejudice your chances of your CV being read or of you getting an interview.

But there are some sentences and phrases that are probably best avoided completely.

Do not begin your letter in a self-promoting way.

Dear Sandra Jones

My proven track record of successfully performing complex analyses on various corporations makes me an ideal candidate for ...

Adjectives which are designed to show how fantastic you are, may simply sound insincere or improbable. Note: below is an example of how _not_ to write about your skills.

My broad experience and range of skills make me a superior candidate for this position.

I also have a wide breadth of invaluable experience of the type that gives you the versatility to place me in a number of contexts with confidence that the level of excellence you expect will be met. ... ... I hope that you'll find my experience, interests, and character intriguing enough to warrant a face-to-face meeting, as I am confident that I could provide value to you and your customers as a member of your team.

So, never overstate or exaggerate your level of expertise and experience.

Other things not worth mentioning are social skills for which you provide no concrete evidence. Never just describe, always demonstrate. The skills below are all unsubstantiated and are therefore _not_ a good example.

In addition to my extensive office experience, I have excellent communication skills. I always maintain a mature, gracious and professional manner when communicating with people, even when difficulties arise.

Some 'experts' recommend that you take the initiative by including at the end of your cover letter a phrase such as:

I will contact you in the next two weeks to see if you require any additional information regarding my qualifications or I will contact you in two weeks to learn more about upcoming employment opportunities with organization.

Source: www.career.vt.edu/JobSearchGuide/CoverLetterSamples.html

For me there are two problems with such phrases:

1. they are totally unneccessary – you can easily make the phone call without having to announce it in advance

2. most HR people don't want to feel that you are too insistent – the HR person wants to feel that <u>he / she</u> is in control of the recruitment process, not the candidate

12.29 Be careful of translating typical letter / email phrases from your own language into English

The typical phrases used when writing letters and emails vary massively from one country to another. This means you need to be very careful when translating such phrases from your own language into English.

For example, the typical ending of a cover letter in your language may be *Waiting for your favorable response* or *I remain in expectation of your rapid reply* but in English these phrases do not exist. Many such phrases are formalities and add no real content to the letter, so the simplest solution is to delete them.

However, a salutation at the beginning and end of a letter or email is normally required. The solution is to use phrases that you are 100% certain exist in English. So in this case the simplest solution is to use *Dear + first name + family name* at the beginning, *and Best regards* at the end. If you adopt this policy, you will not make mistakes. If you are creative or write too much, then you will probably make many mistakes.

You also have to be clear in your mind what information a recruiter will and will not find relevant. For example, in some countries when you graduate from university as a lawyer, accountant, engineer, architect etc, you have to pass an additional state exam which then authorizes you to practise professionally in your chosen field. However, if you are applying for a job outside your own country, providing such information is not only irrelevant but also potentially confusing to people in those countries where such state exams do not operate.

12.30 How reliable are templates for cover letters that I can find on the Internet?

Below is a example of how a PhD student used a template she found on the Internet in order to produce her own cover letter. The only details that I have changed are the dates. The phrases she has copied are in italics. As you will note, she has <u>not</u> used them appropriately.

TEMPLATE	ITALIAN PHD STUDENT'SUSE OF THE CVTIPS TEMPLATE
As website design firms are expanding rapidly due to the increase in small business, the need for web designers is higher than ever.	I have been informed of a job opportunity for a resident in a small animal cardiology at the Health and Life Sciences, School of Veterinary.
Through my past work experience and master's degree program, I have gained the necessary skills to become very successful in the web design industry. *I feel that I would be a valuable asset to your firm* as you continue to expand. My web design and programming background could bring an added level of expertise to your current team.	*Through my past work experience* and my studies formation (DVM, school of specialization in cardiology and actually PhD's in cardiology) and Master's in anesthesia, I have gained the necessary skills to become successful in your resident program. *I feel that I would be a valuable asset to your firm* and for your research team.
I will be finishing my master's degree program in May of 20__, and would be interested in scheduling a meeting with you shortly thereafter. I feel that a meeting would allow us to better understand each others needs.	*I will be finishing my master's degree program in May of 20__* and my PhD in July 20__ *and would be interested in scheduling a meeting with you shortly* there after. *I feel that a meeting would allow us to better understand each others needs.*
Please contact me at your earliest convenience so that we can set up a time that works for the both of us. I am eager to speak with you about the direction that your company is moving in.	*Please contact me at your earliest convenience so that we can set up a time that works for the both of us. I am eager to speak with you about the direction that your University is moving in.*
Thank you for your time and consideration.	*Thank you for your time and consideration.*
Source: http://www.cvtips.com/cover-letter/sending-unsolicited-coverletter.html	

One of the main reasons for non-native English speaking people to use templates is that they hope to avoid making mistakes with their English. However, you cannot assume that just because the template was written by a native speaker, that the content will be appropriate and that the English will be perfect.

Most template letters are written for native speakers, apart from those that appear on EFL and ESL sites (i.e. sites specifically designed for non-natives speakers).

Native speakers are much better able to judge how appropriate the template is for their needs. The template above was intended for people in business in the USA, whereas the PhD student has used the template to apply for a job in research. This means that many of the phrases she has copied are totally inappropriate (e.g. *a valuable asset to your firm* – 'firm' means 'company', she is applying for job in a veterinary hospital!).

Also, phrases such as *I feel that a meeting would allow us to better understand each others needs* and *I am eager to speak with you about the direction that your University is moving in* sound absurd – they are not even, in my opinion, appropriate for a native speaker.

If you indiscriminately lift phrases from a template written for native speakers, the result will be that you letter contains some sentences in perfect English (whether appropriate or not for the context) mixed with your own English (in her second paragraph, the student uses the phrase *studies formation* which means nothing in English).

So, if you do copy phrases, then show your final version to a native speaker to check whether what you have written is appropriate.

In summary, only use templates for the following reasons:

- to learn about how a cover letter is typically laid out
- to get ideas about the possible content

12.31 What impression will hiring managers get if I use a template?

Below is a template intended for a position in business. This template appears on 27 websites offering advice on writing cover letters (I was unable to find the original source). It highlights the dangers of writing a cover letter that is totally generic, i.e. it could have been written by any candidate for practically any business.

Dear Hiring Manager,

I was excited to read about this opening as I have the qualifications you are seeking. I have several years of experience in a wide variety of fields including hi-tech, insurance and the non-profit sector.

Here are some of my skills:

- Verbal and written communications

- Computer proficiency

- Customer service

- Organizing office procedures

In addition to my extensive office experience, I have excellent communication skills. I always maintain a mature, gracious and professional manner when communicating with people, even when difficulties arise.

My broad experience and range of skills make me a superior candidate for this position.

I look forward to hearing from you as soon as possible to arrange a time for an interview.

The problems are:

- the candidate has not taken the trouble to find out the name of the hiring manager

- it is full of generic words, there are only three concrete words *hi-tech, insurance* and *non-profit*

- the skills listed are common to practical anyone who has had some working experience – there is no attempt to show how the candidate is in some way different from the other candidates

- there is no mention of what the candidate knows about the company where he / she is applying for a job

- some parts sound exaggerated and insincere, and thus the candidate loses credibility e.g. the reference to being *excited* and to being a *superior candidate*

- there is an unwarranted presumption that the candidate will be offered an interview

The result is that the hiring manager will be unlikely to read the candidate's CV.

12.32 Will I create a good impression if I use sophisticated grammar and complex sentence constructions?

No.

Look at the example below.

> It is with your organization that I desire to offer marketing, contract administration and project management experience. Having a strong background utilizing a variety of design standards along with proposal requirements, I am certain that my skills and experience, when linked with the vision of your company, will serve to create dramatic, profitable results.

The writer of the example above evidently thought she was going to impress the HR manager with her command of English. But in reality the sentences do not sound natural. Due to their length, they are also hard to read and absorb.

So, as in everything you write, use short sentences in clear English. Your aim is make it easy for the reader to absorb the information you are presenting. The HR manager's impression of the candidate above would be of someone who cannot express herself simply and who is rather insincere.

In addition, don't try to be clever and don't philosophize.

> I'm keen on sports, travelling, arts, especially painting, poetry and music. These are definitely ones of the most impacting aspects on our life, as likely they are incorruptible elements of knowledge to be forwarded to the next generations.

The problems with the above paragraph are:

- his hobbies and interests should be listed in his CV not in a cover letter

- the HR person is not interested in the candidate's opinion of the role of art

- he has tried to show that his interest in art is important, but in doing so he has to use quite complicated concepts and this leads to mistakes. For example, the expressions *impacting aspects* and *incorruptible elements of knowledge* do not exist, and *forwarded* should be *handed down*

12.33 What are the dangers of writing an email cover letter?

Email is a great way to apply for a job. In many cases an employer will actually request submissions via email rather than traditional mail.

You just have to write the email in the same way as you would write a cover letter to be sent via traditional mail. Both should be written with the same high level of attention.

Unfortunately, many job applicants focus on the informality of emails. They thus write their cover letter quickly without even checking it. Here is a real application for a post-doc position.

Hi,

I recently completed the PhD in Information Engineering at the University of Pisa, and I'd like to become a good researcher in this field.

I enclose a copy of my curriculum vitae for your consideration.

I would like to increase my experience and knowledge in your laboratories, because they are very technological and .. technology is fundamental to be a good scientist!

I am avalaible for an interview at any time, and I am avalaible for work immediately.

This email would create a terrible impression on the recipient because:

- it clearly took no more than a couple of minutes to write
- it has no structure
- every sentence begins with *I* so it is completely writer focused: *I can, I am, I have, I need, I want*
- there is no mention of any benefit for the recipient's research team
- it contains mainly redundant information

- the writer has tried to be funny (third paragraph)

- the writer has not checked the spelling (*avalaible* should be *available*)

There is absolutely no impression that the candidate is a serious person who understands the importance of good communication skills.

12.34 What should I if do the recruiter specifically requests not to use a cover letter?

If you are doing an online job application at a recruiter's website then there may be no opportunity to send a cover letter along with your CV. Also, some employees may ask you not to send a cover letter.

This means that some of the information that you would otherwise have put in a cover letter has to be incorporated into your CV. The main missing information is likely to be your demonstration that you have certain soft skills. To learn how to talk about your soft skills within the CV see 9.7.

12.35   So, what does a good cover look like?

On the next page Below is an example of a well laid out and constructed cover / motivational letter.

Center for Economic and Policy Research

1611 Connecticut Avenue

NW, Suite 400

Washington, DC 20009

25 November 2028

Full-time Winter International Program Intern January-May 2029

Dear CEPR Staff,

PARAGRAPH 1 I learnt from your newsletter about this interesting opportunity for an intern. In fact, I have read your web pages on a daily basis since I got to know the CEPR from attending Sally Watson's lecture at the *XVIII Encuentro de economistas internacionales sobre problemas de desarrollo y globalización* last March in La Habana, and it has now become an indispensable resource for my understanding of current social and economic problems.

PARAGRAPH 2 I have spent the last academic year at the *Universidad Nacional Autónoma de México (UNAM)* on an Overseas Exchange Student scholarship from the University of Bologna. In the first semester I attended courses of the Maestría en Economía Política and the Maestría en Estudios Latinoamericanos, whereas I spent my second semester doing research for my postgraduate thesis on the perspectives of the regional integration programme *Alternativa Bolivariana para las Américas (ALBA)*.

PARAGRAPH 3 Because of my past experience as head of a cultural association in Bologna I am used to working in a self-directed group and I perform well on both a personal and institutional level. I also have experience in the organization of international events, due to a long collaboration with the University of Groningen in establishing, running and consolidating the European Commenius Course in Bologna.

PARAGRAPH 4 I believe that the combination of my commitment to learning and researching, my long standing interest in Latin American issues, the skills gained from past work experience and the knowledge of CEPR commitments acquired in these months of passionate reading, will enable me to contribute immediately and directly to the CEPR as an International Program Intern.

Thank you for your time and consideration,

Best regards

Below is an analysis of the above cover / motivation letter.

Layout: Everything is aligned to the left, apart from the subject of the letter which is centered and in bold.

Structure: 1) address 2) date 3) subject line 4) opening salutation 5) four paragraphs 6) closing salutation 7) signature 8) reference to the enclosure

Paragraph 1: a) the candidate says where she learned about the position b) she mentions Sally Watson who presumably will be known to the reader c) she shows appreciation for the work that CEPR is doing.

Paragraph 2: Here she shows how what she has studied fits in perfectly with the CEPR's requirements.

Paragraph 3: The candidate states what she can do and then provides strong evidence of it.

Paragraph 4: Perhaps a little pretentious but in reality it makes the candidate sound very sincere, passionate and committed, and in my opinion is a strong ending to her letter.

Below is an example of a longer cover letter / motivational letter.

I would like to apply for a volunteer position for your "New Volunteering @ ToyHouse Project". Please find attached the application form and my CV.

I am 22 years old, from Pisa (Italy) where I am studying Political Sciences at the University of Pisa. I came to London two years ago, and plan to go back to Italy to finish my degree in June next year.

Currently I'm looking for an opportunity to develop my skills and knowledge in charities and social organizations. The ToyHouse Project appeals to my long-standing interest in childcare and education. In fact, from the age of 15 to 20 I worked as a dance instructor with children from 3 to 12 years of age. It was an amazing working experience that has changed my approach to life and also influenced the choice of my degree. Working with children at such an early age made me really conscious about child labour and how this above all affects developing countries. In addition, during my teenage years I worked at summer camps. My ultimate dream would be to work either in my local community or abroad with NGOs and charities, to help deal with these issues and especially to try to help give these children their childhood back.

I would greatly appreciate the opportunity to be part of your team, and feel sure that your organisation would benefit from my versatile skills. I love spending time with kids and feel that I would be a particularly appropriate person for your Early Years Softplay and Sensory Softplay programs. In addition my fitness training and teaching practice would be appropriate skills for your outdoor Olympic theme program, Hop, Skip & Jump. Furthermore thanks to my experience in the retail sector, I can offer great customer service and help in selecting and stocking toys. Regarding my recent work experience, you will notice from my CV that I have changed jobs quite frequently – each new job has resulted in a higher salary and greater responsibility, and of course, new and useful experiences. I hope you will consider my application because I believe that with my work experience and skills, I would be a positive addition to your team.

I look forward to hearing from you.

Note how the candidate has:

- tried to find the typical things that would be involved in the job and how she would match these needs.

- shown that she is really interested and passionate, and that she has a clear idea of what the job entails. She thus highlights why she is the right person. By doing so she should be able to differentiate herself from all the other applicants.

- mentioned elements from her CV. She has not assumed that the HR person will read her CV in detail

- avoided writing anything that makes it seem that she is exploiting this job opportunity entirely for her own benefit. She makes it look that there will be a clear benefit for her potential employer

12.36 What should I write if I am simply making an enquiry about a possible job (i.e. no job has actually been advertised)?

Look at your chosen company's website. See what positions there are in the company. Find one that matches your qualifications, and then ask if they have such a position free.

You can begin your letter as follows:

I was wondering whether you might have a position available for a software analyst ...

You can then continue as in 12.35.

12.37 My cover letter does not fit on one page, what can I leave out?

In addition to the redundant information mentioned in 12.28, there are other elements of a cover letter that you can easily cut without removing any essential content. Below are some sentences that the candidate of the first cover letter in the previous subsection decided to cut.

I combine a sound academic background and a keen interest in issues related to international development with a strong passion and commitment to the pursuit of economic and social justice through the redefinition of the relationship between economics and politics.

Together with the rest of my personal background presented in the enclosed resume, these experiences have helped me develop good social skills and a capacity for reliable and autonomous working habits.

I would be happy to have the opportunity to experience a period of Internship at the CEPR, assisting the staff in those researches that have been indispensable for my studies and from which I tried to learn the admirable working method.

The problem with the above sentences is that add very little value for the reader. They contain very subjective information or information that is difficult to substantiate. As a result they actually detract from the positive impact of the letter.

So when you have finished your letter, make sure you try to eliminate any sentence that does not serve a very clear purpose i.e. any sentence that will not definitely improve your chances of securing an interview.

12.38   What type of paper should I print my letter on?

Very good quality. The quality of the paper reflects on the quality of you as a candidate.

Summary: Cover letters

➢ From your cover letter the recruiter wants to know:

  ○ what job you are applying for (and perhaps where you saw the advertisement); alternatively, who gave you their name

  ○ why you are interested in this field / company

  ○ how your skills and experience directly apply to the advertised job

  ○ the benefits for the company / institute of employing you

  ○ that you know something about the institute / company

  ○ how you might fit in both in terms of your skills and your personality

➢ Your subject line in the email should include:

  ○ the position you are applying for

  ○ where you saw the advertisement (put this information in brackets)

➢ Structure the letter as follows:

  ○ say what position you want (and where you saw the job advertised)

  ○ say what you're doing now, and when your current position will end

  ○ provide very brief selected past history that will interest the reader and give you credibility

  ○ show that you know about the company or research team and highlight the benefit for them of having you in their team

  ○ brief ending – further details can be given in the next email

➢ Do everything you can (through web searches, via your contacts, or via direct phone contact) to find out the name of the person who is likely to read your CV, this will demonstrate that you really want the job and have a proactive attitude.

➢ Find documents written by the institute / company and imitate their style, layout (e.g. use of white space), font and font size.

➢ Mention your soft skills (ability to work in teams, professionalism, ability to meet deadlines, proactive nature, problem solving, presentation skills, ability to write technical documentation) and briefly provide evidence for them.

➤ Avoid mentioning anything that:

- ○ is not strictly necessary (e.g. your name at the beginning, the fact that you are available for interview and that they can contact you for further information)

- ○ could sound arrogant or negative

- ○ is very subjective and which you provide no evidence for

- ○ makes it sound that the benefit of employing you is solely for you and not for them

➤ Write in clear simple English.

➤ Make sure your cover letter fits on one page. A motivational letter can be longer.

➤ Avoid templates.

13 WRITING A BIO

13.1 What is a bio? When would I need one?

A bio is a biography, i.e. the story of a person's life, or in the context of this book, their academic career. Bios are typically used by people in academia and research, for:

- conferences. If you are giving a presentation at an international conference, the organizers may ask you for a short summary of your career and major achievements. The organizers will then use this bio for the conference proceedings.

- books and book chapters. If you are asked to contribute to a publication, you may be requested to provide a bio.

- your home page, either your personal home page or your institute's home page.

A. Wallwork, *CVs, Resumes, and LinkedIn,*
Guides to Professional English, DOI 10.1007/978-1-4939-0647-5_13,
© Springer Science+Business Media New York 2014

13.2 How do I write a bio for a conference or book chapter?

Bios for both conference proceedings and books tend to be written in a very formal way using the third person (i.e. *he / she, his / her*). They are usually structured as one paragraph.

Typical things to mention include:

1. your degree(s)
2. previous positions
3. your current position
4. what projects you have worked on
5. what project you are working on now
6. your plans for the future
7. the number of first author publications
8. the number of conferences attended where you gave a presentation
9. committees that you are on
10. patents held

Things that you might write on your CV but probably would not include in your bio are:

- non technical skills
- teaching experience
- private interests

Below is an example. The numbers in square brackets correspond to points 1–10 above.

[1] Volmar Thorgaard holds a degree in physics from the University of Copenhagen, Denmark. [2] In 2014 he joined KNUT, an institute of the Danish National Research Council. Starting in early 2016, he spent 18 months at the IBM Scientific Center in Cambridge, Massachusetts, working on computer networks. [4] He has since directed several national and international projects including (in chronological order): xxx, yyy, zzz. [3] In 2018 he joined the Department of Information Engineering of the University of Helsinki, where he is now an assistant professor. [5] His current research interests include the design and performance evaluation of PMAC protocols for wireless networks and quality of service provision in integrated and differentiated services networks. [6] He is planning to bring together Scandinavian countries into a joint project linking up research center computers to Finland's space station on Mars. [7] Thorgaard is the first author of 10 papers published in international journals, [8] and has presented his research at all the major conferences on technologies for Mars in the last decade. [9]

13.2 How do I write a bio for a conference or book chapter? (cont.)

He is on the editorial committee for the two top space technology journals in Scandinavia. [10] He is co-author of several patents with Danish Telecom and Nokia, in the areas of scheduling algorithms.

Note the following:

- Volmar avoids beginning every sentence with *he* by occasionally beginning with a date, using *his*, and using his name (both his full name, and his family name alone)

- present simple to refer to present situations: ***holds*** *a degree in physics,* ***is*** *co-author of several patents*;

- present perfect for situations that began in the past and are still true today (Volmar still directs projects): *he* ***has since directed*** *several national and international projects*

- past simple for finished actions: *In 2018 he* ***joined*** *the Department of Information Engineering*

13.3 How do I write a bio for a home page?

For their personal home page, people generally use the first person (i.e. *I, my*). For the home page of their university or research institute both personal and impersonal forms are used, and the choice depends on how formal you wish to be.

Below is an example from a researcher's personal home page.

> I am a postdoctoral fellow at the European Space Agency and a visitor in physics with Caltech.
>
> My primary research interest is the observation of astrophysically unmodeled bursts of gravitational waves from sources such as core collapse supernovae, the merger of binary compact objects, the progenitors of gamma-ray bursts, or perhaps unanticipated sources. In particular, I am interested in taking advantage of the new global network of interferometric gravitational wave observatories, composed of the LIGO detectors in the US and the GEO and Virgo detectors in Europe, in order to maximize the prospects for detection and the physics that we can learn.
>
> I currently live in Pisa, Italy, and work at the nearby European Gravitational Observatory.
>
> For a list of publications, please see my curriculum vitae (pdf).

Note how she structures her bio into several paragraphs:

1. current position
2. current interests and aspirations
3. indication of where she lives and where she works
4. reference to a pdf version of her CV and list of publications

You also need to give all your contact details.

Note also how she:

- uses the first person pronoun (i.e. *I, my*) but still maintains quite a formal style
- inserts a lot of key technical words – this should increase her chances of being found by a search engine
- is very succinct, she only gives essential information – she doesn't waste the reader's time with unnecessary words or information

Summary: Writing a Bio

Things to mention:

- your academic qualifications
- previous and current positions
- previous, present and future projects
- publications and conferences
- committees, patents and awards

Style

- third person (i.e. *he developed* rather than *I developed*) for conferences and book chapters
- first person for home pages

Be concise, accurate and factual.

14 USING LINKEDIN

14.1 Why do people use LinkedIn?

People create a LinkedIn profile to:

1. boost their chances of finding a job (which is the topic of this chapter)

2. create a network of useful contacts

3. attract clients and business

LinkedIn are constantly updating the look and feel of their website, so this chapter focuses only the elements that are likely to remain constant over time.

14.2 How do recruiters and HR use LinkedIn?

In two ways.

1. Active recruitment. This means they use LinkedIn to find candidates. These are just potential candidates who may not even be on the job market. The recruiter types in keywords (e.g. Master's in xxx, three years' experience in yyy) so that they can find people who they could potentially recruit. They then email these people and present the company to them. Their aim is to arrange an interview with potential candidates.

2. Recruitment via an advertisement. The company or insitute places an ad on LinkedIn and then receives CVs from potential candidates.

A. Wallwork, *CVs, Resumes, and LinkedIn,*
Guides to Professional English, DOI 10.1007/978-1-4939-0647-5_14,
© Springer Science+Business Media New York 2014

14.3 What do potential employers want to see on my LinkedIn page?

First they will probably look at high level details in your profile: who you have worked for and when, and what your job title was (and what this involved), i.e. the same things they look for on a CV. From this information they will get a sense of whether each experience has built upon the previous one and whether you are following some specific direction – again similarly to a CV.

14.4 How does a LinkedIn page differ from a traditional CV?

Your LinkedIn page will differ from your CV in the following areas:

1. Your connections and endorsements: these are things that you cannot build overnight and which may help the recruiter to see:

   - how good your networking skills are

   - who you are connected to (there is a chance the recruiter may have common connections with you)

   - who has endorsed you and for what. However, recruiters are aware that LinkedIn invites members to endorse each other and that such endorsements may be given without much thought

2. People who have recommended you (see Chapter 11 on writing references). If someone has taken the time and trouble to recommend you, this should indicate that they think you made a valuable contribution in some area. But again recruiters are aware that some people write their own recommendations for friends and colleagues to post.

3. Your photo. Although photos are frequently found on CVs (see Chapter 5), they are even more frequently found on LinkedIn as the site encourages you to post your photo.

14.5 I am looking for a job. What key words should I insert and how can I insert them?

In theory, the higher the word count, the more instances of a key word a search engine will find.

Key words tend to be of two types: a) technical, b) skills.

Type A key words are typically nouns or adjectives or combinations of the two (*genetics, genetic engineering; Java, Java script; crytography, crytographic programming*).

Type B key words are often particular skills that a company requires. Such skills are typically written using the *-ing* form of verbs (e.g. *presenting, managing, problem solving*) and adjectives + nouns (software analysis, performance reviews).

Let's imagine you are trying to encourage recruiters and hirers to contact you for a job.

1. On company or institute websites, find 5-10 descriptions of the type of jobs you would be interested in.

2. Underline the key words in each job description. Remember to include both types A and B.

3. Compile a list of the most frequently used 10 keys words.

4. Think of one or two synonyms for each of these words and add these to the list

5. Choose the top 3-5 key words in your list and insert them into your heading.

6. Use as many key words in as many other places in your profile as possible (with the exception of the Personal Interests and Causes sections – see 14.17 and 14.18). Try to insert them in a natural way, so that a human reader will not feel they are being bombarded by key words – remember that you are writing both for search engines and humans.

If your key words can be expressed by several different words, you cannot be sure which of these words a potential employer might use, so you should try to insert all of them. For example, I revise the English of manuscripts for publication in scientific journals. This revision process can be described using several different words – *revising, editing, proofreading, correcting.* So I need to insert all of these in my profile. But I also need to insert different forms of the same key word: *revise, revision, revising.* Then I need to consider the kind of documents I am revising, so I need to include: *paper, manuscript, research, thesis* etc.

14.5 I am looking for a job. What key words should I insert and how can I insert them? (cont.)

I am also a teacher of English. My field exploits many acronyms e.g. EFL, ESL, ESOL, CPE, FCE, BESIG. So it makes sense for me to include both the acronym (BESIG) and the full version (business English special interest group).

Using synonyms and different forms of the same words also makes your profile sound more natural.

Obviously in some cases, only very specific key words will be required, which may have no synonyms. But again, if these words have different grammatical forms (infinitive, gerund, noun, adjective, adverb) you can still create some variety.

To learn how to insert key words see 6.4, 6.8, 8.3 and 8.4.

14.6 How should I write my name and my headline?

To learn how to write your name, see 4.2.

Do not use any titles e.g. *Dr, Professor,* or any qualifications e.g. *MSc, PhD.* Instead simply write your given name and family name.

The headline is the equivalent of an Executive Summary in a CV, see 6.5.

14.7 Can I use the same photograph as on my CV?

Yes, see Chapter 5 to learn how to choose the most appropriate photo.

14.8 Is it OK to use the first person pronoun?

Yes, unlike a CV or bio (see Chapter 13) where some people only use the third person, nearly everyone uses the first person on LinkedIn.

14.9 Do I need to have hundreds of connections?

It will not have a negative impact if you have hundreds of connections. On the other hand only having a few (i.e. less than 100), in the minds of some HR people might make it seem that you are not very proficient at networking.

14.10 What contact info should I put?

Some people will message you directly on LinkedIn, but some may prefer to email you. So provide an email address, which does not have to be your regular address but one you set up specifically to deal with LinkedIn enquiries.

14.11   Should I describe my Experience in the same way as in my CV? What about my Skills and Expertise?

The content is likely to be very similar to your CV (Chapter 8), but don't simply cut and paste sections from your CV. A LinkedIn profile can be considerably longer than a CV, you can afford to use a less telegraphic and more dynamic style. Your aim throughout your profile is to:

*   catch the attention of search engines by using appropriate key words

*   showcase your achievements and make yourself sound unique, i.e. different from other people searching for the same kind of job

*   highlight your credibility

14.12   What words and expressions should I avoid in my profile?

Don't fill your profile with:

*   adjectives that could be interpreted as being an exaggeration (e.g. *above average, amazing, cutting edge, highly*) – you can use these once, but not in every paragraph

*   expressions that are typically found in millions of profiles (e.g. *proven track record, results oriented, team player, hard worker, good communication skills).*

Your aim is to be factual and accurate. No search engine will be programmed to look for such adjectives and expressions. Moreover they add no value for the recruiter, in fact they detract the reader's attention from the important elements in your profile.

14.13   How useful is a video profile?

Most video profiles fail to achieve what the person wants and instead tend to do the opposite, i.e. reflect negatively. For the purposes of getting a job, a video profile, is not necessary. For more on video profiles see 1.13.

14.14   I don't have any Honors & Awards. Is it a problem?

No. You do not need to complete this section.

14.15   Is it important to join groups?

Experts recommend joining the maximum number of groups allowed. But this probably only makes sense if you are running a business, rather than looking for a job.

14.16   What about Volunteer Experience and Causes?

It is safe to talk about volunteer experiences as they imply that you will have learned various social skills. However, listing the causes you support is potentially dangerous – what if the HR person shares very different political views from you?

Even though LinkedIn constantly encourages you to 'complete' your profile, you might consider simply leaving this section blank.

14.17 How should I talk about my interests?

You can talk about your interests in the same way as on a CV (see Chapter 10). To get some ideas for what to include and what not, look on the profiles of your connections and see how they list their interests.

I believe that you should not design this section to attract recruiters or potential clients.

Below is what I have written under my own profile, which I have divided the section into subsections:

Music: ECM jazz, contemporary classical, Terre Thaemlitz, Can, John Martyn, Tricky, Led Zeppelin, Lamb, La Crus

TV series: Borgen, Seinfeld, The Goodwife, Dragon's Den

Etymology: particularly of English and Italian

Fiction: Leo Tolstoy, Ian McEwan, George Eliot, Anna Southern

Sports: hiking, badminton

None of the key words I have written are likely to get me a job or attract new customers to my services. Instead the interests I have listed are targeted at human readers (my students and my editors) who may genuinely want to learn a little more about me as a person rather than me as a teacher or author. I haven't listed all my favorite activities, for example, drinking wine, looking at the incredible view from my house, and walking my dog. I could have included these under a section called 'Other', but I decided that mentioning alcohol may not go down well with my non-drinking students and that the other two points don't reveal anything particularly significant or positive about me. I have also not listed the fact that I take absurd pleasure in reading dictionaries of all kinds from front to back.

14.18 What should I put under 'Advice for contacting'?

You can write anything you want here. Typically you can tell people:

- what you would like to be contacted about (e.g. jobs, projects)
- how you want people to contact you (e.g. your email address)

My section is as follows:

adrian.wallwork@gmail.com

I am interested in writing textbooks in the fields of scientific and business English.

My company is also specialized in revising, editing and proofreading scientific manuscripts written by non-native English researchers.

I also offer courses in how to write and present scientific work.

So I have given my email address and advertised my three skills / services: book writing, editing and English courses.

14.19   How can I get and exploit recommendations?

Recommendations are the LinkedIn equivalent of a reference letter. They serve the same purpose: they provide an objective evaluation of your experience and skills. See Chapter 11 to learn some useful tips on the content and style of a reference letter, which you can also apply with a recommendation.

As with a reference letter, you can:

* ask people to write recommendations for you (11.8, 11.9), for example when you have completed a particular project or provided someone with a service
* write your own recommendations (11.14) and get people to post them on your profile

The main differences from a reference letter are that recommendations:

* are an integral part of your profile (in a CV only the referee is mentioned, the reference letter is a separate document)
* will be scanned by search engines, and thus they should contain your key words
* may be spontaneously offered by past and present colleagues, bosses, and customers

If someone spontaneously writes you a recommendation, you will be automatically asked to authorize the publication of this recommendation. Before you make the authorization, check that the recommendation:

* is written in clear, concise and correct English
* contains key words (8.3-8.4)
* is accurate and factual

You can then ask the person to make any changes that you feel would be appropriate. Remember that your recommendations reflect on you: even if the recommendation is very positive in terms of content, if this content is not presented well it could have a negative impact on your image.

14.20   How often should I update my status?

The experts recommend updating your status daily, but this could be interpreted as you being obsessed with informing the world about your movements and ideas.

14.21　Should I use templates to help me write my profile?

Probably not.

If you do use them, be careful (see Chapter 2 and 12.30 on the dangers of using templates).

A better to solution is to get ideas from the profiles of your contacts.

14.22　What final checks do I need to make?

Paste your profile in a Word file and use the spell checker to ensure that you have not made any spelling mistakes.

As mentioned in 1.8, just one spelling mistake is enough for your CV to be rejected. Likewise, a recruiter will not impressed if they find spelling mistakes in your LinkedIn profile.

Finally, get friends and colleagues to assess the profile for you and to give you critical feedback.

Summary: LinkedIn

Select key words for insertion in your profile, by analysing typical job specifications in your field.

Exploit the full character count in each section by inserting your key words for search engines to capture.

Use the same basic content as in your CV, but present it in more detail and in a slightly more informal and sales-like form.

Avoid words and expressions that add no value. Be wary of using phrases that you have found in templates.

Don't worry about not completing some sections (e.g. Projects, Causes).

Give clear contact details.

15 FINAL THOUGHTS

15.1 How important are my CV, cover letter and other such documents?

Incredibly important.

Some written documents can change the course of your career.

CVs, reference letters, cover letters and LinkedIn profiles are examples of such life-changing documents.

So it is certainly worth paying a professional or at least a native English speaker to correct these three documents. The expense should be minimal, you are only submitting about four pages of text.

However, the benefit is massive.

If your documents contain mistakes in the English this will be a bad reflection on you (particularly if in the Language Skills section of your CV you claim to have advanced or fluent English).

A. Wallwork, *CVs, Resumes, and LinkedIn,*
Guides to Professional English, DOI 10.1007/978-1-4939-0647-5_15,
© Springer Science+Business Media New York 2014

15.2 Am I likely to be a good judge of how accurate, appropriate and effective my CV and cover letter are?

No, you are not - even if you are a native speaker of English.

Quickly skim the cover letter below. On a scale of 0–3, how would you assess the following five factors in the cover letter below?

- English grammar
- English vocabulary
- English expressions
- Structure of the letter
- Relevance of the content

In last decade Nano science and Nanotechnology has been playing a big role in Research and Industrial development. IPR has put the multiplier effect on the scientific and industrial development for socio-economic benefits. IPR reserves the rights of scientific person, industrialist, R&D organization and a common man for their intellectuality, novelty *etc.* From this given background it is clear that I hold keen interest in IPR and thus would like to gain deep knowledge of IPR and its economic effect. It is understood fact that legal expert and scientific personnel are facing lots of problem to design or implement IP issues for Nano science and Nanotechnology work. In this regards I want to explore my skill to get the theoretical as well as practical knowledge.

During my IPR Diploma course, I have found that I am decently skilled in the art of persuasion, as my teachers and my colleague will rightly testify. I've had a knack of getting my point across very well, communicating with people, understanding their needs and providing them with a value proposition which is truly hard to refuse. My skills lie in my ability to comprehensively read and understand the situation and act quickly and yet smartly. But of what use is a raw skill, unless it is sharpened? So to this end, I decided to apply for intensive Summer Course on Intellectual Property and Business Entrepreneurship in IPR at this prestigious organization so as to help me understand more about IPR, to help me understand the mind of the consumer better and to learn some soft skills which have proven to be effective over many years. And laden with textbook knowledge, I wish to implement the skills that I have learned in the real world. I want to prove to myself that I have truly been benefited by this education and what better place to start, than an institution as reputed as yours? For your kind perusal I have enclosed by resume. Deep hope of your encouraging response. Thanks you so much in advance, *please consider my application for WIPO scholarships scheme.*

Unfortunately a recruiter would probably give this candidate a rating of 0 or 1, for each of the items. The English is a strange combination of sentences that could have been written by a native speaker alongside bad mistakes, a mix of colloquial and very formal expressions, and expressions that probably exist in the applicant's native language but do not exist in English e.g. *Deep hope of your encouraging response.*

15.2 Am I likely to be a good judge of how accurate, appropriate and effective my CV and cover letter are? (cont.)

The structure is not conventional. The candidate ends by mentioning the reason for his letter (an application for a scholarship) - this is key information and should be at the beginning. He has highlighted this information by using bold italics, but in reality it would have been better to place it on a separate line at the top of the letter. On the other hand, the beginning reads like an introduction to a scientific paper on the topic of IPR, a topic that presumably the reader will be well acquainted with.

However, the writer was pleased with his letter. He had no idea that it was not a good cover letter. It requires an expert to judge whether your letter is good enough, and I repeat again, it is worth paying someone to revise such an important document.

15.3   How important is my English?

I once received a CV from a non-native teacher of English who wanted to work for me. In the Work Experience section, she wrote that she *teached English at all levels*. To me this indicated that the teacher was incompetent on two levels. She demonstrated that i) she did not have a good command of English, ii) she had not taken the trouble to spell check her CV (given that *teached* is a word that does not exist, a spell checker would have highlighted it).

15.4   Can I use Google Translate to translate my CV and LinkedIn profile?

You can use Google Translate or other such software to provide the first draft, but afterwards you will need to do a lot of work on it. To learn how to use Google Translate see Chapter 21 of my book *English for Academic Correspondence and Socializing* published by Springer.

15.5   How should I label my CV file?

Typically, people think from their own point of view. They thus label their CV to help themselves locate it. So they label their CV like this:

CV generic

CV updated March 2020

CV 2019

But none of the above is very helpful from an HR point of view.

Given that your aim is to facilitate the HR person, label your CV as follows:

first name + second name + CV

15.6 If they contact me for an interview, what should I write back?

Below is a typical email that you might receive from an HR department.

Thank you for sending us your CV. We would like to invite you for an interview on 10 March at 10.00. Please could you confirm that this time would be suitable for you.

Your reply could be:

Thank you for contacting me. I would pleased to come for an interview on 10 March at 10.00.

I very much look forward to meeting you.

If the time is not suitable, then rather than inventing an excuse, you can simply say

Thank you for contacting me. I am very interested in coming for an interview, but unfortunately I cannot attend on 10 March. Would it be possible either the week before or after? I could come at any time of day to suit you.

I apologize for the inconvenience and I very much look forward to meeting you.

If you think it is necessary to explain why you are unable to come on the suggested day, you could write:

Thank you for contacting me. I confirm that I am 100 % interested in coming for an interview, but unfortunately on that day …

… my brother is getting married

… I will still be in Japan on an assignment for my current company.

15.7 What should I do if I receive a rejection letter?

The letter that no one wishes to receive is:

Thank you for sending us your CV. Unfortunately, at the moment we have no suitable positions available.

However, we will keep your CV on our files and should any suitable position arise we will contact you.

Thank you once again for your interest in our company and we would like to take this opportunity to wish you all the best for your future.

There is no point in sending a reply and your CV is probably never going to be looked at again by this company (what they have written about *a suitable position* in the second paragraph is simply a formality).

15.8 What final checks should I make before sending my CV / resume?

Before you send any important document, ask a friend or colleague to check through your final version.

Then:

- check for consistency: have you always used bold, italics and initial capitalization for the same purpose? is your grammar consistent (e.g. when describing your roles have you also used the same grammatical form - developed *three applications for xyz,* wrote *technical documentation for pqr;* rather than a mixture - developing *three applications for xyz,* wrote *technical documentation for pqr,*

- do a very final spelling check. HR people are capable of rejecting candidates simply on the basis that the CV or letter contained one single spelling mistake. Make sure you check the spelling of any names of software, products, institutions, companies etc (i.e. words that an automatic spell checker will not find). Also check for mispellings such as *form* instead of *from,* or *addiction* rather than *addition* i.e. spelling mistakes that an automatic checker cannot find.

Good luck!

15.9 Template for a CV

Below is a possible template for a two-page CV.

For examples of CVs see my website: e4ac.com.

Your name should be in 12 pt, headings in 11 pt and the rest in 10 pt. Your name and personal details should be centered if you have no photo, or aligned to the left with your photo on the right.

The parts in [square brackets] are optional. Obviously, you will have more or less subsections in each section depending on your experience.

Instead of an Executive summary, you may just have an Objective.

To learn more about what to include in each section, see the chapter references in brackets below:

1. name (4)

2. personal details (4)

3. objective / personal statement / executive summary (6)

4. education (7)

5. work experience (8)

6. skills (9)

7. personal interests (10)

8. publications (8.9, 8.10)

9. references (11)

15.9   Template for a CV (cont.)

First Name + Second Name

first.second@email.com; cell phone number

[dd/mm/yyyy; nationality; gender]

Executive Summary

- blah
- blah
- blah
- blah

Work Experience

2026-2032   Name of company + [www.etc]

Position, role + details of work carried out highlighting technical and soft skills

2025-2026   Name of company + [www.etc]

Position, role + details of work carried out highlighting technical and soft skills

Education

2016-2022   Name of university / institute + [www.etc]

Qualification obtained + [further details, highlight technical and soft skills]

2015-2016   Name of university / institute + [www.etc]

Qualification obtained + [further details, , highlight technical and soft skills]

15.9 Template for a CV (cont.)

page 2

Skills

Languages	Language 1: mother tongue; [Other main language: fluent]; English: spoken (proficiency), listening (proficiency), written (proficiency) and reading ((proficiency); [English exams passed: name of exam, grade]
Software	software 1 [level of proficiency]; software 2 [level of proficiency); etc
Technical	technical 1 [level of proficiency]; technical 2 [level of proficiency); etc

Personal interests

Interest 1:	Blah blah blah ...
Interest 2:	Blah blah blah ...
Interest 3:	Blah blah blah ...

Publications

Publication 1

Publication 2 etc

References

Name 1: position; email address; website address

Name 2: position; email address; website address

Name 3: position; email address; website address

15.10 Template for a resume

On the next page is a possible template for a one-page resume.

The parts in [square brackets] are optional. Obviously, you will have more or less subsections in each section depending on your experience.

Instead of an Executive Summary (also called Career Highlights), you may just have an Objective.

Note: In addition to the Experience and Education sections, you may also wish to put one or more of the following sections:

* Associate Memberships / Professional Affiliations
* Certifications
* Honors
* Professional Training
* Publications
* Related Experiences
* Skills (technical and language)

Unlike a CV, a resume generally does not include a photo, a Personal Interest section, or a References section.

15.10 Template for a resume (cont.)

First Name + Second Name

first.second@email.com; cell phone number

Executive Summary

- blah
- blah
- blah
- blah

Experience

Most recent position

Name of company + [www.etc]; dates of employment

5-6 line description of role including key skills (technical and soft)

Second-most recent position

Name of company + [www.etc]; dates of employment

2-4 line description of role including key skills (technical and soft))

etc

Education

Most recent educational qualification

Name of university / institute + [www.etc]; dates of attendance

Qualification obtained + [further details, highlight technical and soft skills]

Second-most recent educational qualification

etc

THE AUTHOR

Adrian Wallwork

I am the author of over 30 books aimed at helping non-native English speakers to communicate more effectively in English. I have published 13 books with Springer Science and Business Media (the publisher of this book), three Business English coursebooks with Oxford University Press, and also other books for Cambridge University Press, Scholastic, and the BBC.

I teach Business English at several IT companies in Pisa (Italy). I also teach PhD students from around the world how to write and present their work in English.

For the last 15 years I have been giving courses to non-native graduates and post graduates on how to write and present their work. These courses include a module on writing CVs and cover letters. In addition, I run my own company (offering teaching and editing services) and have revised countless CVs and job applications over the years.

I have thus collected nearly a thousand CVs written by people of all nationalities, both native and non native speakers of English.

In writing this book I consulted various HR managers both in Europe and the US, as well as recruitment agencies.

CV courses

I run 3-hour seminars on how to write CVs and cover letters. I also hold courses on business English and academic English (for researchers). I am happy to travel anywhere in Europe. To learn more, email me: adrian.wallwork@gmail.com

Contacts and Editing Service

Contact me at: adrian.wallwork@gmail.com

Link up with me at:

www.linkedin.com/pub/dir/Adrian/Wallwork

Learn more about my services at:

e4ac.com

A. Wallwork, *CVs, Resumes, and LinkedIn,*
Guides to Professional English, DOI 10.1007/978-1-4939-0647-5,
© Springer Science+Business Media New York 2014

Index

This index is by section number, not by page number. Numbers in bold refer to whole chapters. Numbers not in bold refer to sections within a chapter.

A. Wallwork, *CVs, Resumes, and LinkedIn,*
Guides to Professional English, DOI 10.1007/978-1-4939-0647-5,
© Springer Science+Business Media New York 2014